Aircraft Dynamics from Modeling to Simulation

Aircraft Dynamics from Modeling to Simulation

Editor

Cezar Dalca

Aircraft Dynamics from Modeling to Simulation

Edited by **Cezar Dalca**

Printed in 2017

ISBN: 978-1-68117-107-4
Library of Congress Control Number: 2015952746

© 2016 by
SCITUS Academics LLC,
616, Corporate Way, Suite 2, 4766,
Valley Cottage, NY 10989

www.scitusacademics.com

Notice

Preface

Aircraft dynamics is the science of air vehicle orientation and control in three dimensions. The three critical flight dynamics parameters are the angles of rotation in three dimensions about the vehicle's center of mass, known as pitch, roll and yaw.Aerospace engineers develop control systems for a vehicle's orientation about its center of mass. The control systems contain actuators, which apply forces in several directions, and generate rotational forces or moments about the aerodynamic center of the aircraft, and thus rotate the aircraft in pitch, roll, or yaw.

Aircraft Dynamics: From Modeling to Simulation provides readers with modern tools for modeling and simulation of aircraft dynamics. The emphasis is on detailed modeling of aerodynamic thrust forces and moments. Topics include aircraft equations of motion, modeling of aerodynamic thrust forces and moments on the aircraft, and analysis of aircraft static and dynamic stability.

This book with specific features for assisting, motivating and engaging aeronautical/aerospace engineering students, in the challenging task of understanding the basic principles of aircraft dynamics and the necessary skills for the modeling of the aerodynamic and thrust forces and moments. Additionally, it also provides a detailed introduction to the development of simple but very effective simulation environments for today demanding students as well as working professionals and researchers.

Table of Contents

CHAPTER 1

Digital virtual flight testing and evaluation method for flight characteristics airworthiness compliance of civil aircraft based on HQRM

Liu Fan, Wang Lixin, Tan Xiangsheng

School of Aeronautic Science and Engineering, Beihang University, Beijing 100191, China

ABSTRACT

In order to incorporate airworthiness requirements for flight characteristics into the entire development cycle of electronic flight control system (EFCS) equipped civil aircraft, digital virtual flight testing and evaluation method based on handling qualities rating method (HQRM) is proposed. First, according to HQRM, flight characteristics airworthiness requirements of civil aircraft in EFCS failure states are determined. On this basis, digital virtual flight testing model, comprising flight task digitized model, pilot controlling model, aircraft motion and atmospheric turbulence model, is used to simulate the realistic process of a pilot controlling an airplane to perform assigned flight tasks. According to the simulation results, flight characteristics airworthiness compliance of the airplane can be evaluated relying on the relevant regulations for handling qualities (HQ) rating. Finally, this method is applied to a type of passenger airplane in a typical EFCS failure state, and preliminary conclusions concerning airworthiness compliance are derived quickly. The research results of this manuscript can provide important theoretical reference for EFCS design and actual airworthiness compliance verification of civil aircraft.

INTRODUCTION

Modern civil aircraft generally uses electronic flight control system (EFCS), which provides an electronic interface between pilot's controls and control surfaces, generating the actual surface commands for control about all three airplane axes. Since the existing airworthiness regulation for transport category aircraft (CCAR-25-R4)[1] only targets at the airplane with mechanical flight control system,[2] for flight characteristics airworthiness compliance verification of civil aircraft in EFCS failure states, the specific condition about handling qualities rating method (HQRM) presented in Appendix 5 of Federal Aviation Administration (FAA) AC 25-7C must be adopted.[2, 3 and 4] At present, flight characteristics airworthiness compliance verification of advanced civil aircraft such as A320, A330, A380, B777 and B787 in EFCS failure states has been completed based on HQRM.[5] In China, the flight characteristics airworthiness compliance verification of ARJ21-700 feeder liner in EFCS failure states will be undertaken for the first time through HQRM.

According to EFCS failure states, atmospheric disturbance levels and flight envelope conditions, HQRM defines the minimum handling qualities (HQ) level of civil aircraft which meets airworthiness requirements. Completion of assigned flight tasks is used to verify whether a civil aircraft meets the requirements of minimum HQ level so as to determine airworthiness compliance of the airplane. The verification method can be flight test or simulator test. Therefore, flight characteristics airworthiness compliance verification based on HQRM is a complex process of multi-factor coupling and pilot-aircraft closed system HQ rating.

By means of digital virtual flight testing technology, airworthiness compliance evaluation is integrated into the whole airplane development cycle, which represents a new concept for modern civil aircraft design.[6, 7, 8, 9 and 10] Since flight characteristics of modern civil aircraft are closely related to the EFCS design, investigations on flight characteristics airworthiness compliance of civil aircraft at early design phases through digital virtual flight testing and evaluation method are of great importance. It can identify the EFCS design defects as soon as possible to avoid high cost associated with modifications at late design stages and to optimize the general aircraft development cycle. Furthermore, preliminary conclusions on flight characteristics airworthiness compliance can be obtained quickly and used to formulate more reasonable and efficient verification schemes for actual flight test and simulator test, to reduce flight risk and to avoid unnecessary verifications. Therefore, benefits such as improving the safety, reducing the cost and shortening the cycle of airworthiness verification can be obtained.

In this article, HQRM is first used to determine the airworthiness requirements for civil aircraft in EFCS failure states. On this basis, digital virtual flight testing and evaluation method for flight characteristics airworthiness compliance of civil aircraft is proposed based on HQRM. This method can be used for a quick preliminary evaluation of flight characteristics airworthiness compliance of civil aircraft in EFCS failure state, and to assist the EFCS design and the actual airworthiness compliance verification of civil aircraft in the whole development cycle.

FLIGHT CHARACTERISTICS AIRWORTHINESS REQUIREMENTS OF CIVIL AIRCRAFT IN EFCS FAILURE STATES

EFCS failure state, atmospheric disturbance level and flight envelope are the major factors that affect controllability and flight safety of civil aircraft. According to HQRM, flight characteristics airworthiness compliance of civil aircraft in EFCS failure states is determined by HQ levels of the airplane under the combinations of the above three factors.

A EFCS failure state, atmospheric disturbance level and flight envelope form a combination event, the occurrence probability of which is generally the product of occurrence probabilities of the three single events, shown as

$$X = X_c X_a X_e \qquad (1)$$

where X is the occurrence probability of the combination event, X_c indicates the occurrence probability of the EFCS failure state, whose value is determined by the EFCS design, X_a is the occurrence probability of the atmospheric disturbance level, whose values of 10^0, 10^{-3} and 10^{-5} represent light, moderate and severe disturbances, respectively, X_e is the occurrence probability of the flight envelope, whose values of 10^0, 10^{-3} and 10^{-5} represent normal, operational and limit flight envelope, respectively. The combination events with the occurrence probability of X less than 10^{-9} (once per 10^9 flight hours) are not required to be carried out airworthiness verification, while those having an occurrence probability of X higher than 10^{-9} and conforming to the practical flight conditions must be verified.

Flight characteristics airworthiness compliance of civil aircraft in EFCS failure states are expressed as follows: for all combination events to be verified, the airplane should at least meet the requirements of minimum HQ level shown in Table 1, where SAT, ADQ and CON stand for satisfactory, adequate and controllable, respectively.

Table 1: Requirements of minimum handling qualities level.

Flight condition $(X_c X_e)$	Atmospheric disturbance level (X_a)								
	Light level			Moderate level			Severe level		
	Normal	Operational	Limit	Normal	Operational	Limit	Normal	Operational	Limit
Probable	SAT	SAT	ADQ	ADQ	CON	CON	CON	CON	CON
Improbable	ADQ	ADQ	CON	CON	CON		CON		

For any combination event, according to the combined probability $X_c X_e$ of the EFCS failure state and the flight envelope, the flight condition under this combination event is "Probable Condition" ($X_c X_e$ in the range of 10^0–10^{-5}) or "Improbable Condition" ($X_c X_e$ in the range of 10^{-5}–10^{-9}). At the same time, in accordance with the atmospheric disturbance level of this combination event, the airworthiness requirements of minimum HQ level can be determined on the basis of Table 1.

As shown in Table 1, HQ level is classified into "Satisfactory", "Adequate" and "Controllable". FAA definition and the corresponding Military Standard lever and quality of each HQ level are shown in Table 2.

Table 2: FAA definition and the corresponding Military Standard lever and quality of each HQ level.

HQ level	FAA definition	Military Standard	
		Level	Quality
Satisfactory	Full performance criteria met with routine pilot effort and attention	1	Satisfactory
Adequate	Adequate for continued safe flight and landing, full or specified reduced performance met, but with heightened pilot effort and attention	2	Acceptable
Controllable	Inadequate for continued safe flight and landing, but controllable for returning to safe flight condition, a safe flight envelope, and /or reconfiguration so that HQ is at least adequate	3	Controllable

For HQ rating by digital virtual flight testing and evaluation method, the FAA definitions of HQ level are on a qualitative basis, while some criteria of Military Standard would be the quantitative basis if necessary. In addition, HQ rating may also refer to some clauses of Airworthiness Regulation CCAR-25-R4.

Therefore, HQ rating basis of civil aircraft in EFCS failure states mainly includes the following four aspects:

(1) The ability to complete the assigned flight tasks. The airplane should be able to complete the assigned flight tasks and meet the relevant performance criteria of Airworthiness Regulation, Military Specification or other standards.
(2) The control and state variables of the airplane. The aircraft control and state variables should comply with various constraints during the performing of flight tasks, such as velocity limit, angle of attack limit and control surface deflection limit.
(3) The control force and control force gradient of the pilot. The pilot control force and control force gradient should meet the requirements of Airworthiness Regulation and Military Specification, such as the maximum control force of wheel in the Clause 25.143(d) of CCAR-25-4R, and the maximum control force gradient for longitudinal maneuvering task in the Clause 3.2.2.2 of MIL-F-8785C.[11]
(4) The pilot controlling model. In order to ensure that digital virtual flight testing fully represents the actual flight test or simulator test, the pilot controlling model should reflect the actual behavioral characteristics of pilot control.

DIGITAL VIRTUAL FLIGHT TESTING AND EVALUATION METHOD FOR AIRWORTHINESS COMPLIANCE

For flight characteristics airworthiness compliance evaluation of civil aircraft in EFCS failure states, first airworthiness requirements should be determined according to HQRM, including all combination events to be verified and the corresponding requirements of minimum HQ level, as described above in Chapter 2. On this basis, for each combination event to be verified, digital virtual flight testing is used to simulate the realistic process of a pilot controlling the airplane to perform the assigned flight tasks, and the airplane actual HQ level is evaluated in terms of the HQ rating basis. Thus, airworthiness compliance conclusion is obtained by comparing the actual HQ level with the minimum HQ level which meets airworthiness requirement.

Therefore, targeting at flight characteristics airworthiness compliance of civil aircraft in EFCS failure states, digital virtual flight testing and evaluation method mainly includes flight task digitized model, pilot controlling model, aircraft motion and atmospheric disturbance model, and digital virtual flight testing and airworthiness compliance evaluation, as shown in Fig. 1.

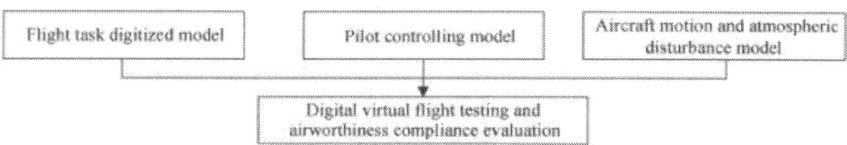

Figure 1: Framework of digital virtual fight testing and evaluation method.

Fight Task Digitized Model

Flight tasks for HQ rating of civil aircraft fall into three categories, as shown in Table 3.[3]"Trim & unattended operation" is mainly to verify the characteristics of an airplane to stay at or depart from an initial trim or unaccelerated condition, which will be appropriate for the failures affecting the trim characteristics of the airplane. "Closed-loop precision tracking of flight path" is generally the pilot closed-loop task performed in routine commercial flight. This type of task will be suitable for the failures affecting the pilot closed-loop tracking performance of the airplane. "Large amplitude maneuvering" basically represents stability and control tests of flight characteristics airworthiness verification, which will apply to the failures affecting stability and control of the airplane.

Table 3: Flight task for HQ rating.[3]

Category	Typical fight task
Trim & unattended operation	Dynamic and flight-path response to pulse (3 axes)
Closed-loop precision tracking of flight path	Crosswind landing
Large amplitude maneuvering	Pitch: symmetric pull-up/push-over
	Roll: rapid bank-to-bank roll
	Yaw: sudden heading change
	Operational: climbing/diving turn

The flight tasks available for choice are not limited to Table 3. In principle, the selected tasks should fully reflect the impact of EFCS failures on the control and flight safety of the airplane under evaluation. Flight task digitized model is to generate digital flight commands based on the requirements of the flight task.[12] Take "Crosswind Landing" as an example, the requirements of this task are written as:[13]

(1) At approach stage, the airplane should glide along the runway centerline at a certain glide angle, and keep a certain velocity.
(2) When the flight altitude is lower than a certain value, the airplane should be flared to enable a mild touchdown at a smaller descent rate.
(3) Before the touchdown, the roll and yaw attitudes of the airplane should be controlled within the safe range.

On this basis, the flight commands for "Crosswind Landing" of a certain passenger airplane are written as

$$
\begin{cases}
V_c = 1.23 V_{SR} \\
\gamma_c = \begin{cases} -3° & h > 6 \\ 0° & h \leqslant 6 \end{cases} \\
y_c = 0 \\
\psi_c = \begin{cases} -\arctan(V_w/V) & h > 3 \\ 0°\text{–}5° & h \leqslant 3 \end{cases}
\end{cases}
\tag{2}
$$

where V_c, γ_c, y_c and ψ_c represent the commands of velocity, glide angle, lateral position and yaw angle, respectively; V_{SR} is the stall speed, V_w the crosswind speed, V the velocity. Eq. (2) does not contain the roll angle command. This is because under the constraint conditions of Eq. (2), the roll angle is determined by the yaw angle.

Pilot Controlling Model

The pilot controlling model is to generate the inputs of throttle and control surfaces according to the flight commands and the feedback of airplane state outputs. The pilot's control behavior of civil aircraft is generally divided into throttle control, longitudinal control, lateral control and directional control, which control the velocity, pitch motion, roll motion and yaw motion of airplane, respectively. For the velocity control, longitudinal control, lateral control and directional control, single-channel pilot controlling model is shown as Fig. 2.[14]

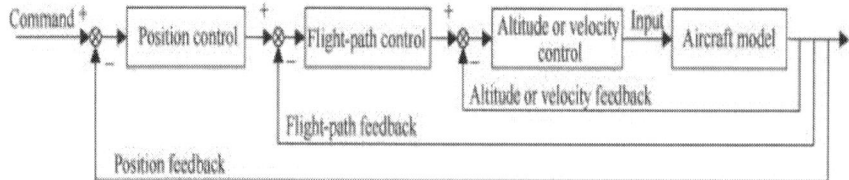

Figure 2: Single-channel pilot controlling model.[14]

As shown in Fig. 2, single-channel pilot controlling mode consists of outer loop for position control, intermediate loop for flight-path control and inner loop for attitude or velocity control. The structure in Fig. 2 is the pilot controlling model with position as flight command. With attitude or velocity control loop retained only, the pilot controlling model with attitude or velocity as flight command is obtained, while with both flight-path control loop and attitude control loop, the pilot controlling model with flight-path as flight command.

For single-channel pilot controlling model, both the outer loop and intermediate loop simply adopt proportional control in order to convert the error signal to the control signal of the inner loop. For the sake of eliminating the steady-state error, the outmost loop can also adopt proportional–integral controller (PI controller). The outmost loop means the intermediate loop when the outer loop does not exist. The inner loop adopts Hess structural pilot model,[15] as shown in Fig. 3, to yield the crossover model of the human pilot.[16] K_m and K_r represent the gains on the output error and the difference $\dot{M}_c - \dot{M}_c$ respectively, and G_{nm} the pilot's limb neuromotor dynamics. Refs. 15 [and] 17 have given the gain scheduling methods for K_m and K_r and presented the expression of G_{nm}.

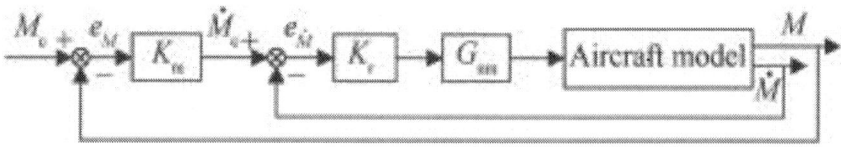

Figure 3: Hess structural pilot model.[15]

The crab method of "Crosswind Landing" is used as an example in the following analysis. According to the flight commands for "Crosswind Landing" shown in Eq. (1), a multi-channel pilot controlling model consisting of throttle channel, pitch channel, roll channel and yaw channel is required to be established to control velocity, slide angle, lateral position and yaw attitude of the airplane, respectively, as shown in Fig. 4, where θ, ϕ, ψ and χ represent the pitch attitude, roll attitude, yaw attitude and head angle, respectively; Y_{PV}, $Y_{P\theta}$, $Y_{P\phi}$ and $Y_{P\psi}$ represent the transfer functions of inner loops of pilot control model in each channel, respectively; $Y_{P\gamma}$ and $Y_{P\chi}$ are the transfer functions of intermediate loop of pilot control model in pitch channel and roll channel, respectively; Y_{Py} is the transfer function of outer loop of pilot control model in roll channel.

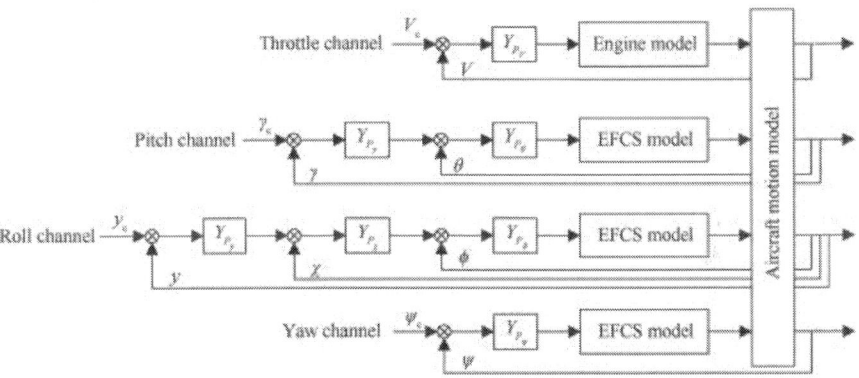

Figure 4: Multi-channel pilot controlling model for "Crosswind Landing".

The throttle channel uses velocity as flight command. The pitch channel uses flight-path angle as flight command, which is responsible for flight-path control loop and pitch attitude control loop. The roll channel uses lateral position as flight command, which is responsible for lateral position control loop, heading control loop and roll attitude control loop. The yaw channel directly uses yaw attitude as flight command.

Aircraft Motion and Atmospheric Disturbance Model
Aircraft Motion Model
Aircraft motion model includes airframe aerodynamic model and EFCS model. In order to build the digital virtual flight testing model based on simulation of "Task–Pilot–Aircraft–Environment", airframe aerodynamic model should be able to reflect the influence of atmospheric disturbance on the motion of the airplane. On the foundation of six-freedom aircraft motion equation in quiet atmosphere,[18] velocity, angle of attack, sideslip angle and aerodynamic moment coefficients are corrected, as shown by Eqs. (3) and (4). Thus, the airframe aerodynamic model in turbulent atmosphere can be established.

$$\begin{cases} \Delta V = -u_w \\ \Delta a = -w_w/V \\ \Delta \beta = -v_w/V \end{cases} \tag{3}$$

$$\begin{cases} \Delta C_l = [C_{lp}(p - \omega_{wy}) + C_{lr}(r - v_{wx})]b/2V \\ \Delta C_m = [C_{mq}(q - \omega_{wx}) + C_{m\dot{a}}(\dot{a}_k - \omega_{wx})]c/2V \\ \Delta C_n = [C_{np}(p - \omega_{wy}) + C_{nr}(r - v_{wx})]c/2V \end{cases} \tag{4}$$

Ref. [19] has given the correction principle and the symbol meanings in Eqs. (3) and (4).

For the EFCS equipped aircraft, the pilot control forces are converted into electric signals, which are sent to the flight control computer (FCC) through wires. Then the FCC performs a calculation based on the pre-configured flight control law and sends electric signals to the control surface actuators. Finally the control surfaces begin to move and the aircraft will achieve the desired flight attitude, flight-path or position. Noteworthy, the FCC can calculate and send electric signals to the control surface actuators directly under the mode of automatic flight control. According to the operation principle of EFCS, the basic EFCS model is shown as Fig. 5.

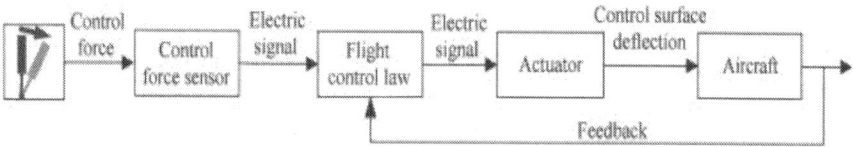

Figure 5: Basic EFCS model.

The control force sensor is the control column or control pedal force sensor model which converts the pilot control force into electric signal. Flight control law is the pre-configured algorithm in FCC. The modeling of control force sensor, flight control law and actuator should be based on the actual EFCS design of civil aircraft.

The failure of EFCS is described by failure mode, failure impact and failure probability.[4] EFCS failures of civil aircraft generally involve pitch, roll and yaw control malfunction, and the specific failure modes include jam, floating, incorrect action, gain variation, rate variation, trim function yaw failure, and so on. The definitions and impacts of typical failure modes are shown in Table 4.

Table 4: Typical failure modes.

Failure mode	Definition	Main impact
Jam	Control column, control pedal or control surface gets stuck in the neutral or normal position	Weakening control effectiveness; generating undesirable force and moment
Floating	Control surface floats to the position where aerodynamic hinge moment keeps zero	
Incorrect action	Control surface acts without instruction	
Gain abnormity	Gain of control surface changes abnormally	Deteriorating handling qualities
Rate abnormity	Control surface acts faster than normal rate	Increasing the sensitivity of pilot control
Trim function failure	Trim function of control surface failures in the neutral or normal position	Heightening pilot effort and attention

The main impacts of EFCS failures are to reduce the control effectiveness and trim capacity of the airplane. In addition, some failures may cause adverse force and moment that impact the airplane control.

According to the flight control system design requirements, the EFCS failures' occurrence probabilities are associated with their impacts, namely, the occurrence probability of failure with severe impact ought to be low. HQRM determines the airworthiness requirements of the minimum HQ level according to the occurrence probability of combination events, as shown in Table 1. Therefore, as a factor of combination event occurrence probability, the EFCS failure occurrence probability will impact on the airworthiness requirements of minimum HQ level, namely, the higher the occurrence probability of EFCS failure, the higher the airworthiness requirements of minimum HQ level.

Atmospheric Disturbance Model
For the flight characteristics airworthiness verification of civil aircraft according to HQRM, atmospheric disturbance required to be considered mainly includes gust, wind shear and crosswind, as shown in Table 5.[3]

Table 5: Atmospheric turbulence for airworthiness verification.[3]

Atmospheric disturbance	Disturbance intensity (m/s)		
	Light level	Moderate level	Severe level
Gust	⩽0.91	0.91–2.44	2.44–6.10
Wind shear	⩽5.14	5.14–12.86	12.86–23.15
Cross wind	⩽5.14	5.14–12.86	12.86–23.15

The model of gust and wind shear[11] are shown in Eqs. (5) and (6), respectively.

$$V_w = \begin{cases} 0 & x < 0 \\ \dfrac{V_m}{2}\left(1 - \cos\dfrac{\pi x}{d_m}\right) & 0 \leqslant x \leqslant d_m \\ V_m & x > d_m \end{cases} \tag{5}$$

$$u_w = u_{20}\frac{\ln(h/Z_0)}{\ln(20/Z_0)}. \tag{6}$$

Ref. [11] has described the model of gust and wind shear in detail and given the meanings of the symbols in Eqs. (5) and (6). Since crosswind is the component of a constant wind that is blowing across the runway, the wind speed can stand for its model.

Digital Virtual Flight Testing and Airworthiness Compliance Evaluation

According to the flight task digitized model, pilot controlling model, aircraft motion model (including airframe aerodynamic model and EFCS model) and atmospheric disturbance model described above, the digital virtual flight testing model is established as shown in Fig. 6.

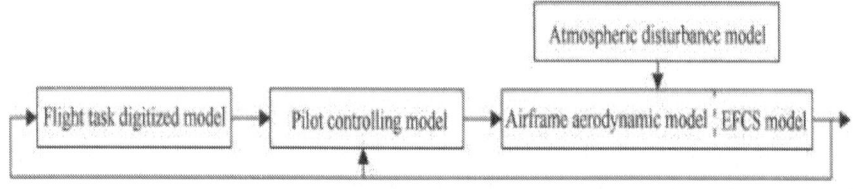

Figure 6: Digital virtual flight testing model.

EXAMPLE   13

For each combination event to be verified, firstly, suitable flight tasks are selected then digitized by the flight task digitized model. After that, the initial flight conditions are set, including flight height, velocity, configuration, center of gravity, state of EFCS of the airplane as well as the wind field, according to the combination event. Finally, the realistic process of pilot performing the flight tasks is facsimiled via the digital virtual flight testing and the relevant data are collected for HQ level evaluation in terms of the HQ rating basis described in Section 2.

For specific combination event, when the evaluating result of HQ level is equal to or better than the corresponding minimum HQ level requirement in Table 1, it is thought that flight characteristics of the airplane under this combination event meet airworthiness requirements.

EXAMPLE

An example of digital virtual flight testing and evaluation for flight characteristics airworthiness compliance of a certain EFCS equipped passenger airplane will be presented. The EFCS failure state is "Control authority degrading of rudder" with the occurrence probability of 5.1×10^{-6}.

"Control authority degrading of rudder" refers to the value of rudder gain at low velocity becoming the value at high velocity. For the example airplane, the rudder gain will decrease with the increasing of velocity. As a result, the deflection range of rudder will reduce due to this failure. ($-20°$ to $20°$) is the maximum deflection range and ($-10°$ to $10°$) is the deflection range at the highest velocity. In order to verify the most serious failure condition, ($-10°$ to $10°$) will be the rudder control authority of the example airplane in this EFCS failure state.

Through calculation and analysis, the airworthiness requirements (including all combination events to be verified and the corresponding requirements of minimum HQ level) and the flight tasks for the example airplane are shown in Table 6.

"Crosswind Landing" is a pilot closed-loop task required to be performed in routine commercial flight and should be examined in normal flight envelope.[3] Therefore, only Cases A and B in Table 6 need to be verified. For "Crosswind Landing", the indexes of HQ rating are shown in Table 7.

Table 6: Airworthiness requirements and flight tasks for the example airplane.

Case	Atmospheric disturbance	Flight envelope	Combination event probability	Flight task	Requirement of minimum HQ level
A	Light	Normal	5.1×10^{-6}	Crosswind Landing	Adequate
B	Moderate	Normal	5.1×10^{-9}	Crosswind Landing	Controllable
C	Light	Operational	5.1×10^{-9}	Crosswind Landing	Adequate

Table 7: Indexes of HQ rating for "Crosswind Landing".

Case	Requirement of minimum HQ level	HQ rating index	Basis
A	Adequate	1. Finishing the task of crosswind landing	Appendix 5 of Ref.3 [and] 20
		2. Both the roll and yaw attitude are less than 6° when touch down	Clause 25.143 of Ref. [1]
		3. The wheel control force and pedal control force is less than 222 N and 667 N, respectively	
B	Controllable	Controllable for returning to safe flight condition, a safe flight envelope, and/or reconfiguration so that HQ is at least adequate	Appendix 5 of Ref. [3]

According to the characteristics of the example airplane, the flight commands for "Crosswind Landing" are shown as Eq. (2); the pilot controlling model is shown as Fig. 4; the initial flight conditions of digital virtual flight testing are shown in Table 8.

Table 8: Initial flight conditions for "Crosswind Landing".

Case	Configuration	Crosswind speed (m/s)	Velocity	Altitude (m)	Weight condition (adverse)
A	Landing	−5.14	$1.23V_{SR}$	100	Maximum landing weight, maximum inertia moment, forward center of gravity
B	Approaching	−12.86	$1.23V_{SR}$	100	

EXAMPLE   15

By means of digital virtual flight testing, the time histories of control and state variables of the example airplane performing "Crosswind Landing" are shown in Fig. 7 and Fig. 8, respectively. Since the failure "control authority degrading of rudder" primarily influences the lateral–directional control of the example airplane, the simulation results mainly include the lateral-directional control and state variables are given.

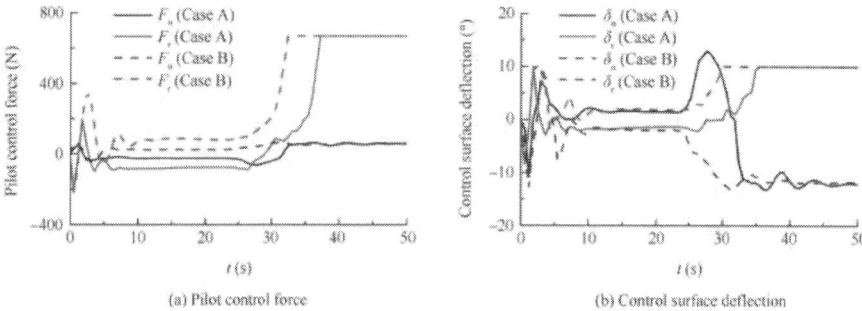

(a) Pilot control force

(b) Control surface deflection

Figure 7: Control variables of "Crosswind Landing".

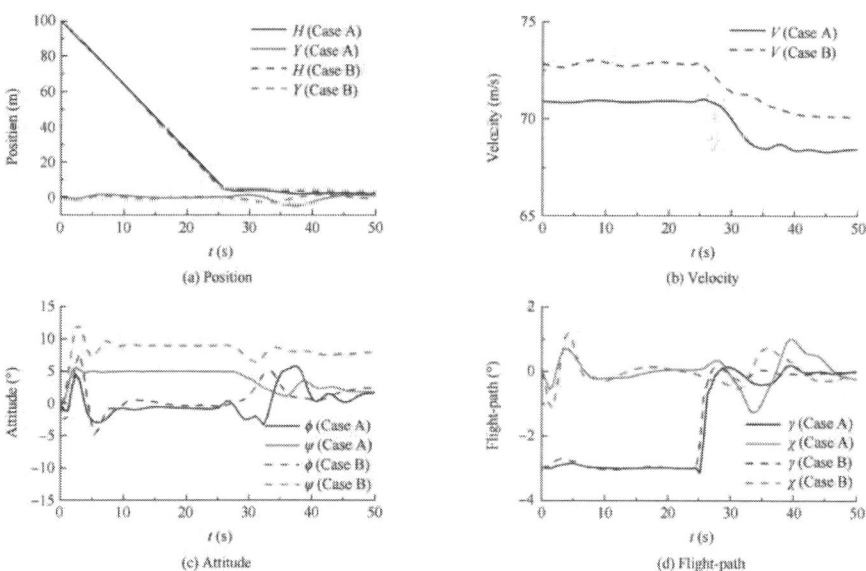

(a) Position

(b) Velocity

(c) Attitude

(d) Flight-path

Figure 8: State variables of "Crosswind Landing".

As shown in Fig. 7, F_a and F_r represent the control forces of wheel and pedal, respectively, δ_a and δ_r are the deflections of aileron and rudder, respectively. As shown in Fig. 8, H, Y and V represent the flight altitude, lateral position and velocity.

From the simulation curve, it can be seen that the whole flight process is divided into three stages: initial disturbed motion, steady approach and flaring before touchdown.

Since the failure does not produce any force and moment, the example airplane is only subjected to the crosswind disturbance at the initial stage: the airplane rolls to left under the action of the right crosswind and the heading deviates from the runway centerline (where $Y = 0$) and drifts left (the initial 2 s in Fig. 8(a), (c) and (d)). In order to resist the disturbance by crosswind, the pilot has to apply upwind ailerons to control the airplane to roll to the right, thus to eliminate the heading deviation; the upwind rudder is simultaneously applied to control the aircraft to yaw to right, therefore to turn the nose pointing to the direction of incoming flow (the 2–5 s in Fig. 7, the airplane maintains a steady approach during the next 20 s, and keeps the heading along with the runway centerline (see Fig. 8(d)). When the flight altitude is lower than 6 m, the pilot flatters out the airplane to reduce the descent rate (see Fig. 8(a)). When the flight height is lower than 3 m, downwind rudder is applied to reducing the yaw angle, with a simultaneous application of upwind ailerons to counteract the airplane left rolling trend caused by the right crosswind (the last 25 s in Fig. 7). Finally, the example airplane is grounded near the runway centerline (see Fig. 8(a)).

According to the terms of the HQ rating basis described in Chapter 2. First, as shown inFig. 8, in both Cases A and B, the example airplane can finish the flight task of "Crosswind Landing". Second, the control and state variables of the airplane comply with the constraints, except for the yaw angle exceeding the safety range upon touchdown in Case B, as shown in Fig. 8(c) and Table 7. Third, for the Cases A and B, during the whole flight process, the wheel control forces are less than 167 N, and the pedal forces are less than 667 N (see Fig. 7(a)), which meet the requirements on control force shown in Table 7. Finally, the selection of pilot controlling model gains are carried out according to Ref.[17], which can reflect the actual behavioral characteristics of the pilot. Pilot controlling model gains are shown in Table 9, where K_{p1} and K_{i1} represent the proportional gain and integral gain of PI controller in outer loop, respectively; K_{p2} and K_{i2} the proportional gain and integral gain of PI controller in intermediate loop, respectively. As shown in Table 9, the element G_{nm} is a second-order system model.

Table 9: Pilot controlling model gains.

Channel	Outer loop		Intermediate loop		Inner loop		
	K_{p1}	K_{i1}	K_{p2}	K_{i2}	K_m	K_r	G_{nm}
Throttle					0.62	0.2	$\dfrac{100}{s^2 + 2 \times 0.707 \times 10 \times s + 100}$
Pitch			1.5	1	2.12	2.36	
Roll	0.004	0.0013	5.3		1.98	0.52	
Yaw					1.72	−5.22	

Therefore, the airplane actual HQ in Case A acquires level of "sufficient", which meets the airworthiness requirement. The airplane actual HQ in Case B does not acquire level of "controllable" and fails to meet the airworthiness requirement.

According to the digital virtual flight testing and evaluation results above, suggestions about the design and the actual airworthiness compliance verification of the example airplane are proposed and stated as follows.

(1) Increase the rudder size of the example airplane appropriately to make up for the diminution of yaw control effectiveness due to the failure.
(2) Reduce the occurrence probability of "Control authority degrading of rudder" by improving the system reliability of the EFCS to avoid the occurrence of Case B.
(3) The digital virtual flight testing and evaluation results can provide references for the verification scheme formulation and the pilot's controlling strategy identification of the actual flight test and simulator test in the verification process.

CONCLUSIONS

(1) For flight characteristics airworthiness compliance of civil aircraft in EFCS failure states, digital virtual flight testing and evaluation method is proposed based on HQRM. First, we determine the airworthiness requirements for flight characteristics and HQ rating basis for airworthiness compliance evaluation according to HQRM. On this basis, we build a digital virtual flight testing model for simulating the realistic process of a pilot controlling an airplane to perform assigned flight tasks. Finally, we rate the HQ level and evaluate the airworthiness compliance according to simulation results and HQ rating basis.

(2) The research results of this manuscript can provide valuable theoretical references for EFCS design and actual airworthiness compliance verification of civil aircraft: airworthiness requirements can be incorporated into the entire development cycle of civil aircraft, and preliminary conclusions on airworthiness compliance can be obtained quickly, which is valuable for verifying and improving the EFCS design, formulating efficient scheme of the actual flight test and simulator test, as well as identifying riskless pilot controlling strategy in the verification process.

ACKNOWLEDGEMENTS

The authors are grateful to Dr. Xu Dongsong for joint research. They also thank the anonymous reviewers for their critical and constructive review of the manuscript. This study was supported by the National High-tech Research and Development Program of China (No. 2014AA110500).

REFERENCES

1. Civil Aviation Administration of China. China civil aviation regulations part 25 airworthiness standards of transport category aircraft. Beijing: Civil Aviation Administration of China; 2011. p. 1–28 [Chinese].
2. Zhu L, Huang MY, Ou XP, Song ZT. Analysis of key technologies of airworthiness certification of electronic flight control systems for civil transport airplane. CGNCC 2010: 2010 Chinese guidance, navigation, and control conference. Shanghai: Science Press; 2010 [Chinese].
3. Federal Aviation Administration. Flight test guide for certification of transport category airplanes. Washington, DC: Federal Aviation Administration; 2012. p. A5-1–10.
4. Zhang YN, Li Y, Jin L. Airworthiness verification of handling qualities of the aircraft with electronic flight control system. Flight Dyn 2012;30(2):117–20 [Chinese].
5. Zhang YN, Li Y, Gao YK. Commercial transport aircraft flight simulation flying qualities airworthiness verification. Appl Mech Mater 2012;235:170–5.
6. Mendonca CB, Silva ET, Curvo M, Trabasso LG. Model-based flight testing. J Airc 2013;50(1):176–86.
7. Xu HJ, Liu DL, Xue Y, Zhou L, Min GL. Airworthiness compliance verification method based on simulation of complex system. Chin J Aeronaut 2012;25(5):681–90.

8. Schari J, Mavris DN, Burdun IY. Use of flight simulation in early design: formulation and application of the virtual testing and evaluation methodology. 2000. Report No.: AIAA-2000-5590.

9. Burdun IY, DeLaurentis DA, Mavris DN. Modeling and simulation of airworthiness requirements for an HSCT prototype in early design. 1998. Report No.: AIAA-1998-4936.

10. Hines DO. Simulation in support of flight testing. 2000. Report No.: RTO-AG-300.

11. U.S. Department of Defense. Flight qualities of piloted aircraft. Arlington: U.S. Department of Defense; 1997. p. 679–84.

12. Robert KH. Use of a task-pilot-vehicle (TPV) model as a tool for flight simulator math model development; 2010. Report No.: AIAA-2010-7620.

13. Federal Aviation Administration. Airplane flying handbook. Washington, DC: Federal Aviation Administration; 2004. p. 813–5.

14. Hosman R, Schuring J, Geest P. Pilot model development for the manual balked landing maneuver. 2005. Report No.: AIAA-2005- 5884.

15. Hess RA. Simplified approach for modeling pilot pursuit control behavior in multi-loop flight control tasks. J Aerosp Eng 2006; 220(G2):85–102.

16. McRuer DT, Jex HR. A review of quasi-linear pilot models. IEEE Trans Hum Factors Electron 1967;8(3):231–49.

17. Hess RA. Obtaining multi-loop pursuit-control pilot models from computer simulation. 2007. Report No.: AIAA-2007-247.

18. Fang ZP, Chen WC, Zhang SG. Aircraft flight dynamics. 1st ed. Beijing: Beihang University Press; 2005. p. 174–81 [Chinese].

19. Xiao YL, Jin CJ. The principle of flight in atmospheric disturbance. 1st ed. Beijing: National Defense Industry Press; 2002. p. 82–4 [Chinese].

20. Airbus. Flight operations briefing note landing techniques [Internet]. 2008 Mar [cited 2014 Feb 21]. Available from: http://www.airbus.com/fileadmin/media_gallery/files/safety_library_items/AirbusSafetyLib_-FLT_OPS-LAND-SEQ05.pdf.

CITATION

Fan Liu, Lixin Wang, Xiangsheng Tan, Digital virtual flight testing and evaluation method for flight characteristics airworthiness compliance of civil aircraft based on HQRM, Chinese Journal of Aeronautics, Volume 28, Issue 1, February 2015, Pages 112-120, ISSN 1000-9361, http://dx.doi.org/10.1016/j.cja.2014.12.013.

CHAPTER 2

Cold Expansion Technology of Connection Holes in Aircraft Structures: A Review and Prospect

Fu Yucan, Ge Ende, Su Honghua, Xu Jiuhua, Li Renzheng

College of Mechanical and Electrical Engineering, Nanjing University of Aeronautics and Astronautics, Nanjing 210016, China

ABSTRACT

As one kind of key anti-fatigue manufacture approaches with simplicity and effectiveness, the hole cold expansion technology satisfies the increasing needs for light weight and durability of aircraft structures. It can improve the fatigue life by several times at no additional weight conditions. The hole cold expansion technology has been widely used in manufacturing and repairing of both fighters and commercial aircraft, and has become a research hotspot in the strengthening technology. In recent years, hole cold expansion process methods, residual stress around expanded holes, the behavior of fatigue crack initiation and propagation, and fatigue lives after cold expansion are researched extensively through lots of experiments and finite element simulations. A review on the hole cold expansion technology research status in the last twenty years is presented in this paper. Via the analysis of the current characteristics and defects of the hole cold expansion technology, combined with the actual needs in design and manufacture of new-generation aircraft, development trends and novel research directions are presented for realizing precise and high-efficiency anti-fatigue manufacture.

INTRODUCTION

Light weight and durability have become the common goals of design and manufacturing engineers, for both advanced fighter and modern commercial aircraft. Fatigue loading may lead to failures of structures and components even when the load level is much lower than the ultimate strengths of materials, causing serious consequences.[1] Fatigue failure becomes one of the most important failure modes in aircraft structures.[2] In order to improve the fatigue life of aircraft structures, a lot of research on two aspects, that is, advanced materials and manufacturing processes, has been done by scholars and aviation manufacturing enterprises. On one hand, novel titanium alloys and composites as representatives of advanced materials are widely used in modern aircraft structures, reducing the overall weights of aircraft structures and improving the fatigue resistance performances of structures during service periods. On the other hand, using structure optimization and different anti-fatigue manufacturing process methods, the fatigue life of aircraft is improved and the service time is prolonged effectively.

In aircraft structures, three types of joints are used, namely, mechanically fastened joints, adhesively bonded joints, and welded joints. With the special requirements on safety and durability, mechanical connection including bolted and riveted is still the dominant connection mechanism in primary aircraft structural components.[3] However, implementing riveted or bolted joints needs the components to be drilled to create fastener holes, which causes geometrical discontinuities and entail local stress (or strain) concentration during loading.[4] Investigations showed that there were a number of accidents caused by fatigue failures initiating from fastened joints. According to statistics, fatigue fractures of fasten holes account for 50%–90% of fractures of aging planes.[5 and 6]

In order to countervail the unfavorable effect of holes on the fatigue life of notched components, various techniques for improving fatigue life have been proposed, such as laser shock, shot peening, cold expansion, and interference fitting.[7 and 8] The appropriate interference fit is an effective process method for improving the fatigue life of bolted joints. However, the interference degree is difficult to precisely control. The large amount of interference is easy to damage the hole surface, and the protuberance produced in installation is harmful to the strength of joints.[9 and 10] It not only affects the normal assembly process but also reduces the connection strength. The shot peening process basically modifies a surface and the characterizing parameters of the near surface substrate including principally elastic residual stress distribution, increased surface hardness, higher near-surface dislocation density, and alternated surface topography.[11]Nevertheless, over high surface roughness caused by shot peening offsets the contribution to prolong the fatigue life by residual stress, and reduces the assembly stiffness of an aircraft structure. Laser shock processing is a competitive technology as a kind of green method of imparting compressive residual stresses to improve fatigue and corrosion properties.[12 and 13] Unfortunately, costly high power laser, high operation cost, and instability of the production process limit the application of laser shock processing in aircraft assembly.

A widely used technique for cold expansion fastener holes has been developed by Boeing Company in the early 1970s and was firstly used in F/A-18 and other aircraft structure components.[14] In the cold expansion technique (the term "coldworking" was used by the inventor and early investigators), an oversized tapered mandrel goes through a hole. During this process, elastic–plastic deformation is generated and then compressive residual stress is produced around the hole. This residual stress can reduce the concentration stress and delay the initiation and propagation of fatigue cracks. It is highly effective in preventing premature fatigue failure undergoing cyclic loading.[15] The technique has been applied to critical holes in highly loaded zones of structures, such as landing gears and engine mounting regions. The cold expansion technique can be not only used for new aircraft designing, but also applied to repairing of in-service aircraft. With further studies of the cold expansion technology, a lot of improvements on the cold expansion process have been made in the aviation industry, and novel cold expansion approaches have been proposed and researched in recent years.

It is well known that aluminium alloys have been used as the primary material for structural components of aircraft for more than 80 years,[16] and the current research of cold expansion still focuses on these aviation aluminium alloys. For the "one generation material one generation aircraft" aviation manufacturing industry, the hole expansion anti-fatigue manufacture technology is relatively backward. Especially since entering the new century, new materials have been widely used in aircraft structures, such as new titanium alloys, aluminum lithium alloys, and composite materials. It needs to carry out cold expansion research about various materials in order to maximize the fatigue life. In addition, current studies are limited to single-plate strengthening. Research for the cold expansion technology of multi-plate assembly structure has not been publicly reported. Because of the great influence of external environment on the strengthening effect, it is necessary that the hole cold expansion technology in different external loading conditions should be researched extensively. Furthermore, the development of the cold expansion technology is promoted by the major projects of large aircraft.

In this paper, a review is given on the current status of cold expansion processes, containing process methods, strengthening mechanisms, reinforcement results, and influence factors. Base on that, some disadvantages of current research are analyzed and then corresponding solution countermeasures are pointed out. In light of the current demands of light weight and durability for modern aircraft with complex structures and new materials, the development trends and difficulties urgently needed to be solved for the cold expansion technology are addressed. Finally, five novel research directions of the cold expansion technology are presented and prospected.

APPLICATION OF THE COLD EXPANSION TECHNOLOGY IN THE AVIATION INDUSTRY

Generally, fatigue failure is the most prevalent mode of failure in aircraft structures. While there are many fatigue critical locations and components of aircraft, one common aircraft fatigue location is the fastener hole, especially in the main bearing component, for example, the connection holes of a wing and a fuselage. The cold expansion technology, a kind of simple and effective anti-fatigue manufacturing technology, has been widely used in key fastening holes of various aircraft. Table 1 lists the application of the cold expansion process in some early aircraft types.[8] At the same time, cold expansion were applied in the girder holes of MIG aircraft in Soviet aviation industry, such as MiG-17, MiG-19, MiG-21, etc. The damage tolerance design philosophy has been used in modern aircraft, and the fatigue life of components has been paid more attention in the aviation industry. The cold expansion technology is not only used in manufacturing and repairing for traditional aircraft structures, but also applied in durability design of new model aircraft.

Table 1: Application of the cold expansion process in some aircraft types.

Strengthening method	Aircraft type	Strengthening position
Cold expansion process	F-16, F-18, B-707, B-747, B-757, B-767, MD-82, F-15, Kfir, etc.	Main bearing components
Cold expansion process and interference fit bolting	J-5, DC-9, DC-10, etc.	Lower surface of central wing, etc.
Cold expansion process and interference fit riveting	B-707, etc.	Connection of stiffening profile and skin in the wing box spanner torque wall. Connection of stringer and skin in the lower wing panel. Connection of fuselage frame and skin

INVESTIGATION OF THE COLD EXPANSION PROCESS

Over the past decades, the cold expansion technology has been used extensively in the aircraft industry to improve the fatigue life of mechanical connections structures with fastener holes. Many cold expansion approaches have been developed and investigated according to the diverse expansion tools. The most widely used methods in practice are four cold expansion techniques: hole edge expansion, direct mandrel expansion (without sleeve), ball expansion, and split sleeve expansion processes.

Hole edge Expansion

The hole edge expansion process is carried out by means of a high-hardness tapered indenter or a rigid ball to extrude or hammer the hole edge under the action of axial force F. This process method is shown in Fig. 1. During the hole edge expansion process, plastic deformation occurs in the material near the hole edge, and then a zone of compressive residual stress is generated around the hole edge. However, due to the middle part of the hole wall is not directly strengthened, the fatigue gain effect brought by this approach is limited. Therefore, the hole edge expansion process is not suitable for thick-plate connecting holes, and only used for strengthening of holes in sheet metal parts. In order to enhance the fatigue strength of parts, it usually needs to carry out multiple expansions for holes. [17] Liu et al. [18] investigated the effect of the hammer peening process on the fatigue behavior of holes in aluminium alloy 2A12-T4, and found that the fatigue life of these hammer-peened holes was significantly prolonged fivefold compared to non-treated holes. For skin parts, considering the aerodynamic requirements of aircraft, the dimple treatment is necessary for assembly holes. The experiment and analysis results show that the hole edge expansion process is better for sheet metal with holes effectively.

Direct Mandrel Expansion

The direct mandrel expansion process is one of the earliest developed methods of anti-fatigue manufacture in the aviation industry, for example, Douglas Company. In this way, the cold expansion process consists of pushing a pre-lubricated tapered mandrel through the hole from the entrance side of a specimen and removing it from the other side. Fig. 2 illustrates the mandrel direct expansion method. The hole is expanded to an extent sufficient enough to cause permanent plastic deformation. Upon the removal of the mandrel, the surrounding elastic material attempts to return to its undeformed state, producing a proper distribution of compressive residual stresses around the

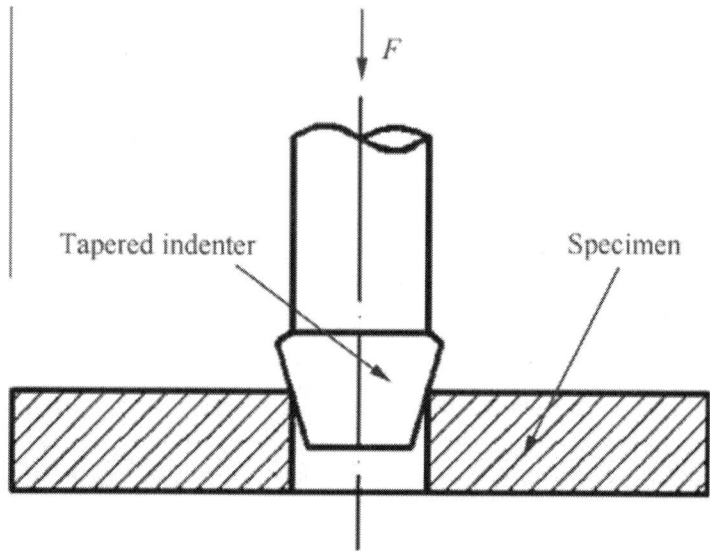

Figure: 1: Hole edge expansion process.

fastener hole. This residual stress can delay the initiation and propagation of fatigue cracks, and contribute to improving fatigue life. Furthermore, due to the direct contact between the mandrel and the hole surface, this method has advantages of burnishing and smoothing the hole surface.

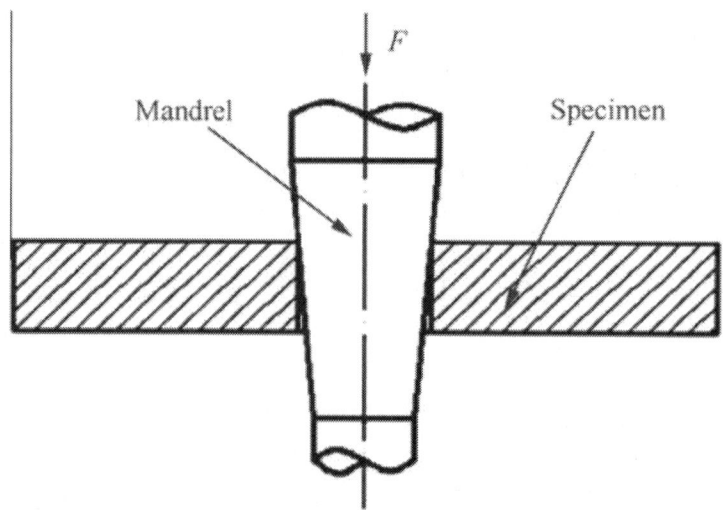

Figure 2: Direct mandrel expansion process.

The strengthening mechanisms and strengthening effects were researched extensively.19, 20, 21, 22 and 23 because the direct mandrel expansion method is easy in operation, it has been widely used in the anti-fatigue manufacture of aircraft structural parts.

Ball Expansion

The ball expansion process is carried out by inserting a pre-lubricated and oversized hard steel ball from one side of the holed plate, and then followed by the removal of the ball from the other side.24 and 25 The expansion process is depicted in Fig. 3. Due to producing a localized interference ring between the steel ball and the hole surface, the friction is smaller in comparison with the mandrel expansion method. Therefore, this method may be used in anti-fatigue manufacture of small holes on alloy steel plates.

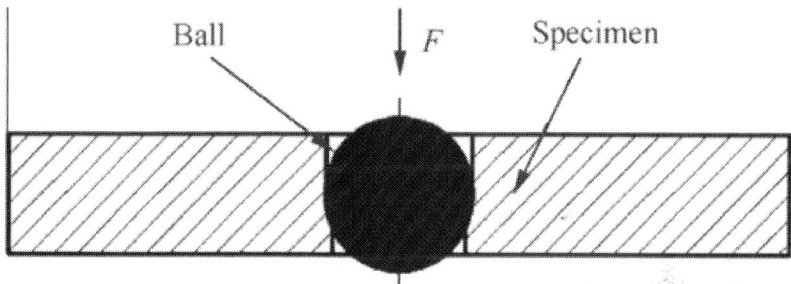

Figure 3: Ball expansion process.

Due to producing a localized residual tensile ring at the entry side of the hole, the beneficial effect on fatigue life from this method is smaller in comparison with other expansion methods. In order to overcome this problem, a double ball expansion process is investigated. In addition, the area of the compressive residual stress zone is larger in the hole surface, in the sense that the hole drilling followed by double expansion can give rise to a significant increase in the number of cycles for the initiation of fatigue cracks, in comparison with a single ball expansion.[26]

Split Sleeve Expansion

A major impediment to the use of a ball or a mandrel in the cold hole expansion of fasteners is the surface damage introduced at the interface during the cold expansion process. To overcome this difficulty, the split sleeve expansion process is used. A schematic drawing of this process is shown in Fig. 4. In this technique, a solid tapered mandrel and an internally lubricated stainless steel split sleeve are utilized. The split sleeve is placed over the mandrel, and the mandrel/sleeve assembly is then inserted into an accurately sized hole. Plastic deformation of the material is generated when the part of the mandrel with a larger diameter moves through the sleeve. When the mandrel is withdrawn from the sleeve, some elastic recovery can take place, and the split sleeve is removed from the hole after cold expansion, leaving a permanent enlargement of the hole and a desired compressive residual stress.[27] Because of the existence of the split in the sleeve, a small raised pip is formed on the bore of the hole surface. A reaming operation to size the hole accurately is carried out after cold working which removes the pip and also avoids the occurrence of cracks near the pip.[28]

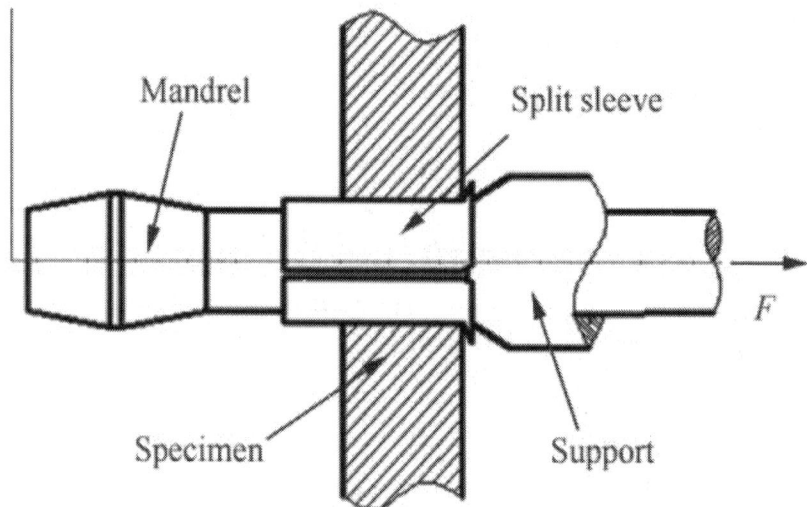

Figure 4: Split sleeve expansion process.

This expansion method has the advantages of good adaptability, high production efficiency, and small damage around the hole surface after expansion. Comparing with other existing cold expansion techniques, the split sleeve cold expansion method has been more widely used in the aerospace industry[29, 30 and 31], especially in the bad openness connection of aircraft assembly.

RESIDUAL STRESS INVESTIGATION

The local plastic deformation strengthening process is one of the strengthening methods of active application of residual stress. The cold expansion technology just utilizes this residual stress caused by plastic deformation to achieve anti-fatigue manufacture. The strengthening effect is affected by the size and distribution of residual stress. Therefore, the investigation of the residual stress field is the focus of attention for the cold expansion strengthening technology.

Residual Stress Measurement Method

Since it is the compressive residual stress field produced by cold expansion which is responsible for the amelioration of stress concentration at faster holes, accurate measurement of these fields has been the focus of many experimental programs. In the early studies, researchers have made a careful analysis of the residual stress around an expansion hole, starting with material deformation and springback in the theory of elastoplasticity.[32 and 33] It is important to note, however, that due to the limitation of measurement techniques, stress was not directly measureable in the initial researches. With the progress of means of measurement, various test methods are used to study the residual stress after cold expansion. According to the measurement area, it may be divided into two types, namely, point measurement and face measurement. Point measurement methods, for example, the strain gage method,[31] the hole-drilling method, the X-ray diffraction method,[34, 35 and 36] single beam laser speckle interferometry,[37] etc., have been widely applied in the measurement of the residual stress in an annular region adjacent to the hole. As a distinguishing feature of these methods, the residual stress field is fitted with some point stress by repeated measurement. Nevertheless, it should be pointed out that the residual stress cannot be accurately obtained at a point in the region close to the hole since the experimental measurements are averaged over a 1 mm × 1 mm area. To overcome this problem, face measurement methods such as moiré interferometry[38 and 39] and the photoelastic coating method[40, 41 and 42] are utilized. Although the mechanism and accuracy of these measurement methods are different from each other, the typical distribution of residual stress measured by various methods around a cold expanded hole is basically similar. The typical distribution of tangential residual stresses σ_θ around a cold expanded circular hole is shown in Fig. 5.

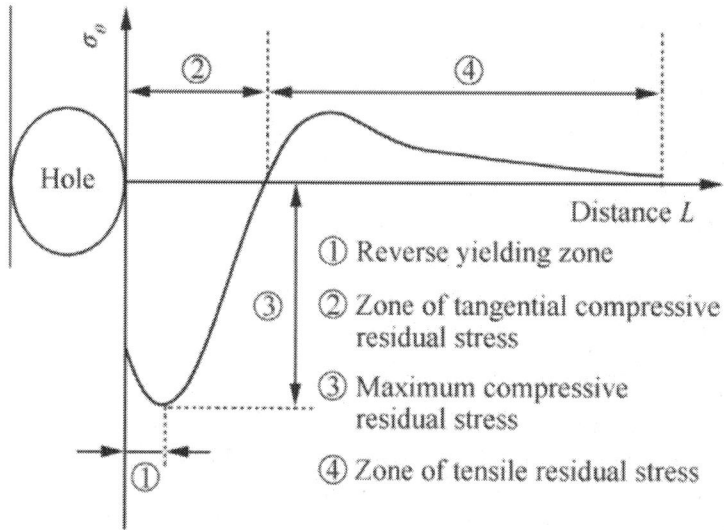

Figure 5: Typical tangential residual stress distribution around an expanded hole.

Residual Stress Simulation Research

Although many analytical models and experimental techniques viz. X-ray diffraction, single-beam laser speckle interferometry, photoelastic coating, etc., have been employed and developed to predict the residual stresses field, it is still difficult to capture the three-dimensional distribution of the residual stress near the hole wall surface, due to inherent limitations of different experimental methods. The elastic–plastic deformation zone of a metal material is only a small part around the cold expanded hole, so the compressive residual stress is limited to a certain small areas around the hole. Because of the limitations of measurement range and accuracy of existing test methods, the residual stress field with a large gradient variation still cannot be measured accurately. In other words, it is difficult to measure the stress field of the internal material near the hole wall. With the development of computers, finite element simulation can be widely applied to study the residual stress generated by the cold expansion process. Through the rational designs of meshes, material properties, contact conditions, boundary conditions, etc., accurate residual stress distributions can be obtained from numerical simulation.

Various analytical models and numerical simulations have been tried and used to investigate the residual stress. In the early 1990s, two-dimensional axisymmetric finite element (FE) approximation models have been designed for the cold expansion of a fastener hole.[28 and 35] The tangential and radial residual stresses are obtained around the cold expanded hole. It is shown that there is good agreement between simulation and experimental results. Although some limited numerical works have been carried out to consider the residual stress along the thickness of a sheet metal, the two-dimensional finite element simulation models of cold expansion are not adequate to accurately estimate the residual stress field resulted from the cold expansion process, due to the asymmetry of the residual stress field through the thickness of the plate induced by the cold working process.[43]

In recent years, three-dimensional finite element simulation models have been utilized to research cold expansion.[44 and 45] A simplified 3-D finite element simulation of the cold expansion of a fastener hole has been developed by employing a new approach by Mahendra et al.[46] and this approach is capable of capturing the variation of residual stresses at various sections along the thickness. In order to obtain a more accurate residual stress distribution around the hole, it would be good to set out the prerequisites for the creation of an adequate FE model. It has been proven that there are two key moments in the building of an adequate finite element model: the first is modeling of a realistic contact mandrel-workpiece with or without a mediator and a workpiece-support, and the second is a suitable constitutive model of the workpiece material.[47] Finite element analysis results show that, during the hole cold expansion process, the residual compressive stress of the hole edge exhibits grades across the thickness, with the least amount of residual compressive stress at the tool entrance face, the largest amount of residual compressive stress in the mid-bore, and good residual compressive stress at the tool exit face. The simulation results were consistent with those of experiments by the X-ray diffraction method.[34] Artificial neural networks (ANN) simulation modeling was built and trained to simulate the stress topography surrounding an expanded hole by Toktas and Özdemir,[48] and the result showed that diffraction methods and current ANN predictions were overlaid and similarities in residual stress distributions perceived to valid only at regions where the strain gradient was not changing precipitously.

Influencing Factors of Residual Stress

The stress fields near a connection hole have a great influence on fatigue life under alternate loads. In order to improve the fatigue life, it is necessary that the factors affecting residual stress should be researched after hole cold expansion. It has been found that predicted residual stresses are sensitive to the details of the process, particularly the expansion degree and the geometry relation.

A variety of methods mentioned above can generate the residual stress around the cold expansion hole. Research shows that the reaming of the material around the hole has a slight effect on the maximum value and distribution of residual stresses.[45] Process parameters have great influences on the residual stress, such as mandrel speed,[49]lubrication, expansion degree, etc. From the experimental and simulation results, it is understood that the residual stress increased with the expansion degree.[50] and 51Because of the different material properties, the correlation between process parameters and the expansion degree is not the same. The geometry relation is a crucial factor in the design of aircraft strength. The main dimensions of a part with a hole is shown in Fig. 6, including h (thickness), d (hole diameter), e (the distance from the hole center to the specimen boundary along the width direction), and then the edge distance ratio is e/d. The effects of key processing and design parameters have been systematically studied.[52, 53, 54, 55 and 56] A small e/d results in a change of the residual stress distribution around a cold expanded hole, and it influences the fatigue life. [57] When e/d is more than 3, it has no effect on the residual stress field. [54]

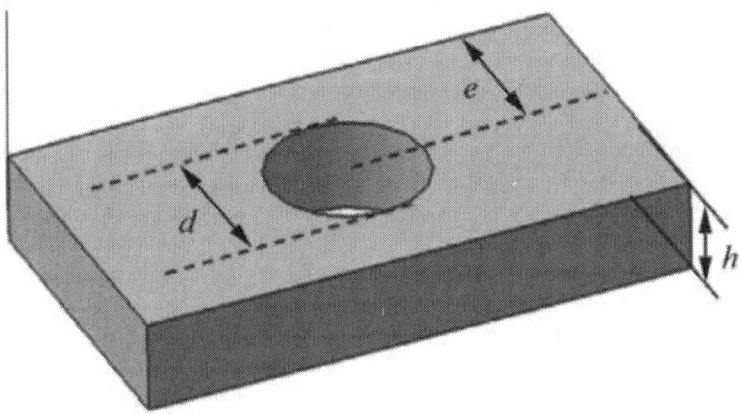

Figure 6: Main dimensions of a part with a hole.

FATIGUE CRACK INVESTIGATION

The fatigue fracture process is divided into three stages, including fatigue crack initiation, propagation, and fracture.[58] Through the investigation of the behaviors of crack initiation and propagation of a hole with cold expansion strengthening treatment, the internal relationship between fatigue life and residual stress can be established. Furthermore, the fatigue strengthening mechanism of cold expansion can be revealed.

Fatigue Crack Source

Fatigue damage often occurs in a stress concentration position, for example, the specimen surface or defect, or a hole. Results suggest that all specimens with a hole broke at the plane of the smallest cross section, as shown in Fig. 7.[59]

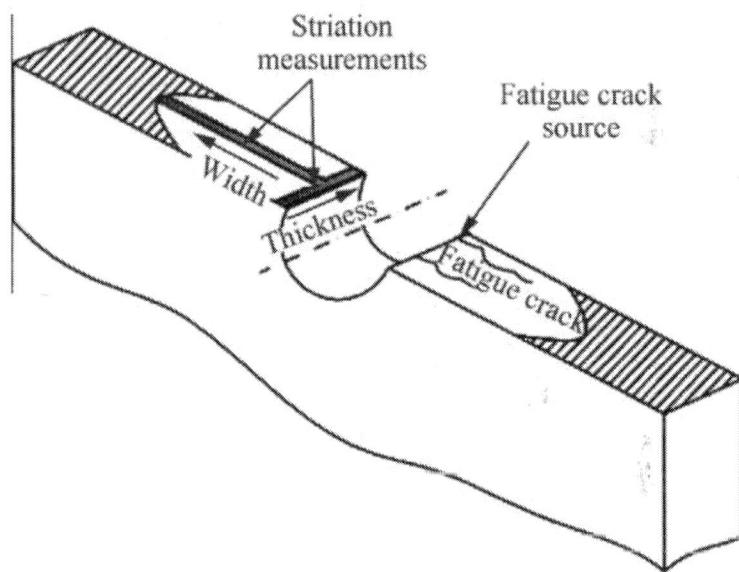

Figure 7: Fractography analysis of a metal fatigue specimen.[59]

As mentioned previously, the maximum compressive tangential residual stress occurs at the hole edge near the mid-plane, whereas the smallest compressive residual stress occurs at the entrance face.[46] SEM studies indicate that, in cold expanded fatigue specimens, fatigue cracks initiated and propagated at the mandrel entrance face around the hole edge, and in unexpanded specimens, fatigue cracks initiated at the corner of the hole and the surface of the hole bore, respectively; the former have only a single fatigue source while the latter have multiple sources.[60, 61, 62, 63] and [64] In cold expanded holes, the propagation speed of fatigue cracks along the transverse direction is faster than that along the axial direction.[65] The distribution of residual stress and the regulations of fatigue crack initiation and propagation are in good agreement.

Fatigue Crack Propagation

Crack morphology reflects the whole process of fatigue crack initiation and propagation. In the fatigue life, the crack propagation life is longer than the crack initiation life. In order to increase the fatigue life effectively, it needs to extend the crack propagation time as long as possible using anti-fatigue manufacture approaches. Studies on crack propagation mainly include the size of the crack propagation area, crack propagation rate, stress intensity factor, etc.

Under the same condition of the crack propagation rate, enlarging the size of the crack propagation area can prolong the fatigue propagation life. Because of the differences of cold expansion processes and mechanical properties, the size of the fatigue crack propagation area is different from each other after hole cold expansion. Studies suggest that the size of the crack propagation area increases with the increasing of the expansion degree.[66] The fatigue striation which is perpendicular to the crack propagation direction can be seen clearly in the fracture surface. The rates of fatigue crack propagation are determined by the measurement of striation spacing.[67 and 68] It has been found that the cowrie pattern spacing resulted from cold expanded holes is always smaller than that resulted from "as drilled" holes, which indicates that the cold expansion process retards the crack propagation rate.[62] In order to research the fatigue crack growth rate, the stress intensity factor must be known and it has been widely studied recently.[69, 70, 71 and 72] Due to the complexity of residual stress distributions around holes, the stress intensity factor is difficult to be used directly for description of the fatigue life. The superposition principle has been used to account for the presence of residual stresses near cold expanded holes. In fact, the effect of residual stress may be explained using a fatigue crack closure model in which compressive residual stress reduces the effective stress intensity factor, consequently reducing the crack growth rate.[73 and 74] The effective stress intensity factor (K_{eff}) is equal to the residual stress intensity factor (K_{res}) plus the stress intensity factor caused by the external applied load (K_{app}). In order to estimate the residual stress intensity factor along the crack plane, the closure-based method and the weight function technique are used.[75] The effective stress intensity factor can easily solve the problem of crack propagation, and a certain relationship between the fatigue life and the residual stress field is established via this factor. It is helpful to the study of the fatigue life which is strengthened by the cold expansion technology.

FATIGUE LIFE INVESTIGATION

Anti-fatigue Effect

Anti-fatigue manufacture is the fundamental purpose of the hole expansion process. The fatigue gain effect has been studied extensively in recent years. According to the differences of materials and expansion process methods and parameters, the gain coefficient is also different from each other. In general, the fatigue lives of aluminium alloys are increased more than those of other metals, because they have good ductility. Research results show that the fatigue life of 7050 high-strength aluminium alloy can be increased 6 times using the hole cold expansion process, while that of 7B50-T7451 can be increased 28 times.[16, 76 and 77] However, the gain coefficients of low-plasticity metal materials are relatively smaller, for example, the fatigue lives of Ti6Al4V and 30CrMnSiNi2A can be increased 2.8 times and 1.7 times.[61 and 62] In addition, stress levels also affect the gain coefficient. For example, in the same conditions, the fatigue life improvement factor is about 3.2 and 1.5 for a degree of cold expansion of 5.58% when applying nominal stresses of 191 MPa and 300 MPa, respectively.[78]

Fatigue Life Prediction

Fatigue life usually consists of the initiation life and the propagation life of a fatigue crack. Fatigue crack initiation and propagation life estimations have been investigated extensively and the total estimated fatigue lives have been compared with available experimental fatigue test results. The cold expansion process can generate compressive residual stress near the expanded holes. The superposition of the mean stress between the residual stress and the external stress is used to analyze fatigue crack initiation. The ε-N method is implemented to predict the fatigue crack initiation life.[79] Another method is put forward to predict the fatigue crack initiation life of LY12CZ alloy after hole cold expansion under variable amplitude loadings and has been substantiated by a fatigue test under a flight-to-flight spectrum.[80] Based on fatigue experiments, fracture analysis, and residual stress computed by elasto-plastic finite element simulation for cold expanded fastener holes, an engineering approach is used to predict fatigue life, and the result is well agreed with that of the experiment.[81] A reliability analysis method is presented and used to analyze the effects of different tolerances and strengthening parameters on the fatigue life distributions.[82] Fatigue life estimation can be used to determine the optimum degree of cold expansion avoiding long-time fatigue tests.

DEVELOPMENT CONCEPTION OF COLD EXPANSION TECHNOLOGY

With the tireless efforts of researchers and engineers, cold expansion technology has been applied in the vast majority of military and commercial aircraft. It makes sense in reducing the stress concentration around fastener holes, and also improving the fatigue lives of aircraft components. However, the high request to the manufacturing and design of aircraft is put forward in recent years. On one hand, a large number of new materials are used for the primary structure in the fourth-generation fighters and civil commercial aircraft. The current cold expansion process cannot be implemented directly in the new materials and structures of aircraft. It is necessary to carry out research of a novel cold expansion technology so as to meet the requirement in practice. On the other hand, the beneficial effect of the cold expansion process is the focus of attention for aircraft designers. In order to accurately predict the fatigue life, the influence of environmental factors on cold expanded components should be researched in the aircraft service conditions. Unfortunately, research on this aspect is not carried out extensively. Numerical simulation has been carried out in cold expansion researches; however, it is still confined to the residual stress around expanded holes. The study on life estimation for cold expanded structures should be developed by synthesis simulation technology. According to the above analysis, five research directions of cold expansion technology are proposed as shown in Fig. 8.

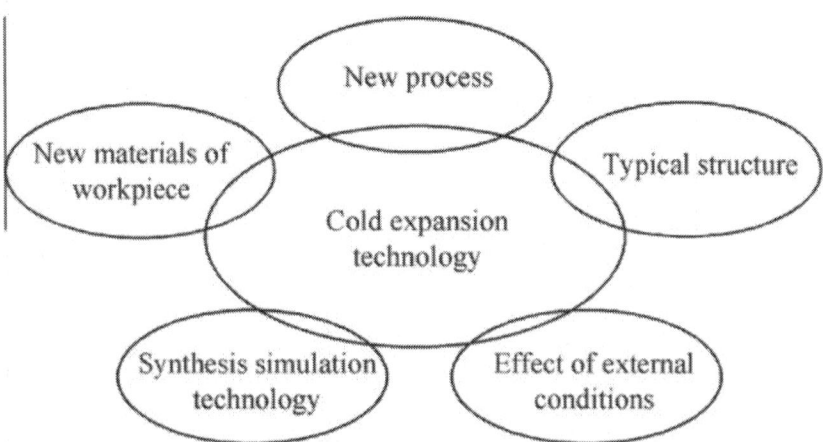

Figure 8: Research directions of cold expansion technology.

Synthesis Simulation Technology

Finite element simulation has been widely applied in research of hole cold expansion technology, especially analysis of the residual stress around expanded holes. In view of the different cold expansion process approaches, such as direct mandrel expansion and split sleeve cold expansion, various residual stress fields are obtained via finite element simulation. Experiment results demonstrate that the accuracy of the residual stress field can be improved through optimizing the geometric model, boundary conditions and mesh, etc. Researches of the residual stress and the crack propagation factor reveal the strengthening mechanism of hole cold expansion technology.

In current studies, however, the simulations of fatigue crack initiation and propagation are not carried out, and the relationship between the fatigue life and the cold expansion process is not established by finite element simulation. It is necessary to carry out synthesis simulation technology which includes the residual stress, the fatigue properties of materials, and the load spectrum, as shown in Fig. 9. First of all, the relationship between the residual stress and the fatigue crack should be established by researching on the mechanism of crack initiation and propagation near the expanded holes, which is a key step towards the cold expansion fatigue life simulation. Via fatigue analyses about the fatigue properties of materials, the load spectrum, and the simulation results of the residual stress, the fatigue life can be estimated with specific cold expansion process and environmental conditions.

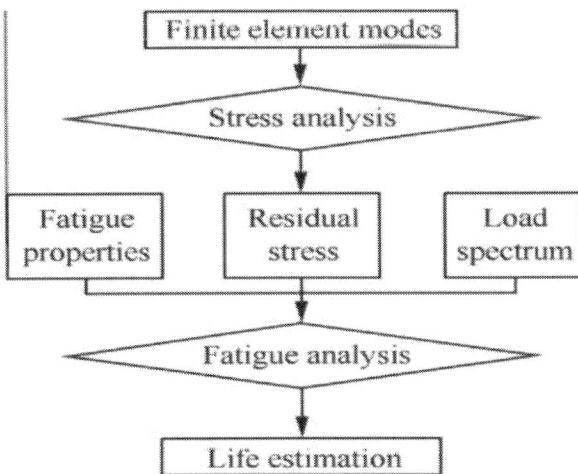

Figure 9: Synthesis finite element simulation for fatigue life estimation of cold expanded specimens.

New Process

Despite the cold expansion technology has been widely used in the assembly and repair of aircraft, the conventional approaches usually have certain limitations.[83] On one hand, cold expansion without sleeves is easy to cause hole surface and mandrel damages, and it is difficult to carry out a large degree of cold expansion. On the other hand, the disadvantage faced by the split sleeve expansion process is necessary reaming operation; moreover, each split sleeve can only be used for processing one hole. It not only causes the complexity of the technology, but also increases the cost of production.

To further improve the performance of cold expansion, new advanced processes have attracted more focus currently. Double cold expansion (ball cold expansion method) can realize relatively large expansion degrees, and the second expansion is carried out in the opposite direction of the first expansion employing a second ball with a larger diameter.[26] A method which allows the calculation of the optimal radial interference and the optimal mandrel shape on a cold-expanded bushing is introduced in order to obtain the desired values of the residual stresses on the hole surface.[84] The surface upset is an undesirable effect which accompanies all mandrel expansion methods; however, the fretting fatigue phenomenon is strongly provoked by the surface upset. With the purpose of reducing the surface upset, Maximov et al.[85] and [86] presented a symmetric cold expansion method and tool (shown in Fig. 10) ensuring "pure" radial cold hole expansion – introducing nearly uniform residual hoop stresses around the hole along its axis having a minimized and symmetric gradient with respect to the plate middle plane, and another advantage of the method is that it ensures one-sided processes like the split sleeve and split mandrel. In addition, according to the actual impact factors including material, geometric dimension, application environment, etc., it is necessary to design suitable cold expansion process approaches and parameters to maximize the fatigue lives of components.

New Materials of Workpiece

The development of the aviation industry is always accompanied by the development of new materials. The hole cold expansion techniques applied to the new materials need to be researched and developed according to their physical and mechanical properties respectively.

New Metals

Titanium alloys are widely used in the new-generation large aircraft, both military transports and commercial airliners, as shown in Fig. 11.[87] According to different positions, there are various titanium alloys with

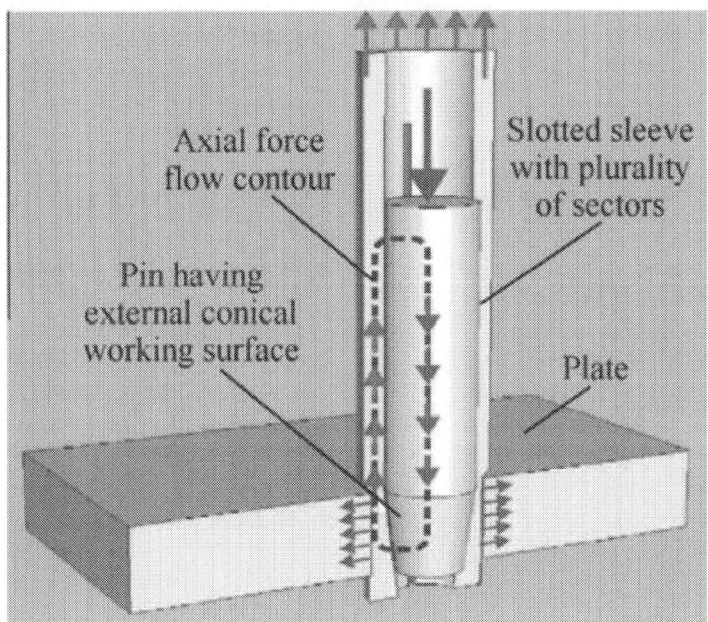

Figure 10: Scheme of symmetric cold expansion.[85]

different properties, especially in the main load-bearing components, for example, engine, keel beam, wing, fin, etc. Compared with aluminium alloys, titanium alloys have high strength and toughness, but poor cold expansion shaping. As an available anti-fatigue manufacture process, however, the cold expansion technology of titanium alloys is barely referred in current documents.[57 and 62]

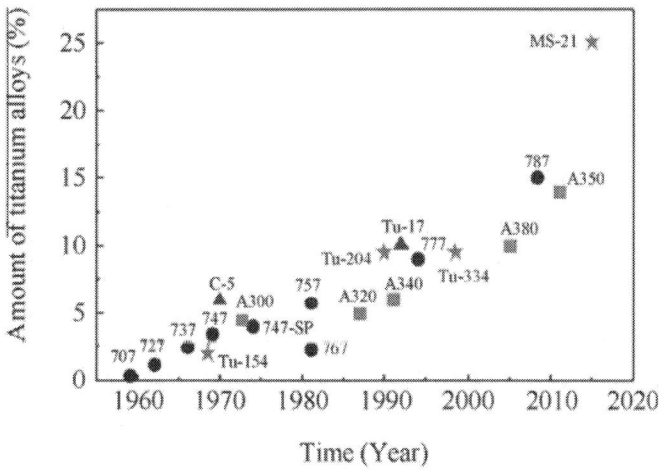

Figure 11: Change of titanium application in large airliners and military transports.[87]

In addition to titanium alloys, other new alloy materials have been widely used in aircraft. The high hot strength and significant resistance to heat and fatigue of nickel-based superalloy lead to its wide application in turbine sections of jet engines.[88] However, the excellent performance also results in its poor deformation ability, especially the cold expansion deformation of structural parts. According to experimental results, it is shown that the alloy GH169 has a strengthened layer with a nearly 3-mm depth from the hole edge.[89] In the aspect of light alloy with low density, the third-generation Al–Li alloys and Al–Mg alloys have been developed and applied in modern airplane structures.[16, 90 and 91]

According to public documents, hole cold expansion still focuses on the traditional materials – various types of aluminium alloys. Because of the different material properties, the cold expansion process for aluminium alloy holes cannot be directly applied to strengthening treatment of new metal materials. Combined with the application conditions and the physical and mechanical properties of metal materials respectively, cold expansion strengthening and anti-fatigue manufacture should be further investigated.

Composite

Due to their low weights and high mechanical properties, nearly for a decade, composite materials have been more widely used in aeronautical and aerospace structures for the fuselage and the wing as well as other structural parts (shown in Fig. 12[87]). The amount of fiber composites applied in commercial airplanes becomes an important index to evaluate the advanced technology of the aviation industry.

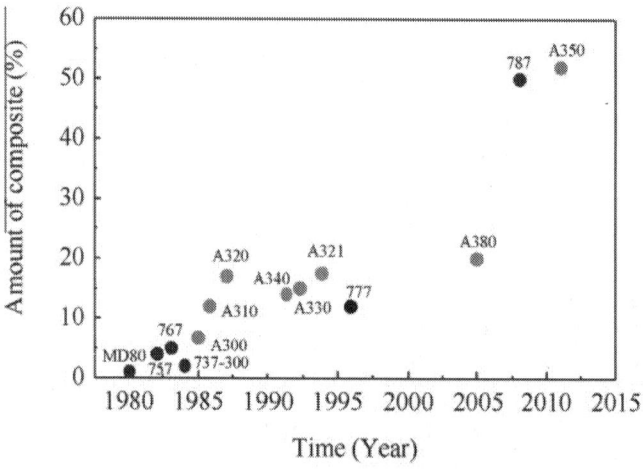

Figure 12: Change of composite application in large airliners.[87]

Anti-fatigue connection technology of composite materials has become an important research topic in aircraft manufacturing. In general, three types of joints are used in composite structures, namely, mechanically fastened joints, adhesively bonded joints, and hybrid mechanically fastened/adhesively bonded joints.[92] Mechanical connection is the main connection way for aircraft parts. Bolted joints are still the dominant fastening mechanisms used in jointing of primary structural parts.[93] Sometimes, as many as 55,000 holes are generally required to be drilled in the complete single unit production of an Airbus A350 aircraft.[94] However, the presence of connection holes compromises the structure integrity, and causes a local stress or strain rise in aircraft parts. On the other hand, drilling composite joint components, to make bolt holes, may create fiber pullouts or delamination in the hole surface. The stress concentration and damage around holes reduce the connection strength, and decrease the fatigue lives of aircraft parts under the action of alternating loads. Compared with metallic material structures, fastener holes are the weak link in composite material structures. The fatigue fractures of fastener holes account for 60%–80% of structure fractures of aircraft.[95]

From the cold expansion technology of metal holes, Fatigue Technologies Incorporated (FTI) has developed a cold expansion process with a metal sleeve for composite holes in aircraft assembly structures (shown in Fig. 13). However, the specific information of this approach is still a secret. In order to improve the fatigue lives of composite material structures, the cold expansion technology for composite holes has become an urgent requirement of aircraft manufacturing enterprises. Therefore, the strengthening mechanism, process, and effect of cold expansion composite holes need further analysis and reveal through a large number of experimental researches.

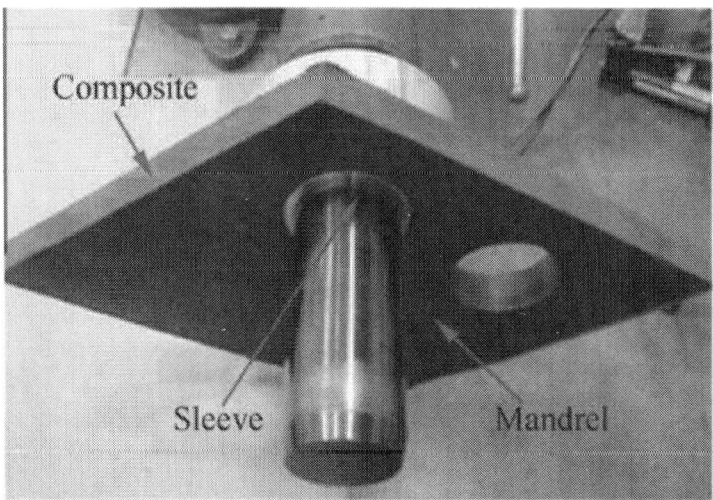

Figure 13: Cold expansion process for a composite hole.

Typical Structure

Although cold expansion technology has been studied for decades, it still focuses on the hole strengthening of a single plate. Actually, various components are assembled through multi-hole connection in the aircraft manufacture. The service life of aircraft is determined by assembly structures. Therefore, it is very necessary to research the cold expansion of typical structures for improving the fatigue life of aircraft.

Multi-Hole Structure

In a mechanical connection structure, the load-bearing components are assembled by numerous fasteners. Particularly in aircraft, there exist a lot of fastener holes at regular intervals. A typical part is shown in Fig. 14. Due to the complexity of the structure, the research of stress distribution and fatigue damage about these holes is relatively difficult. In that case, the residual stress variation might be influenced by the adjacent holes if cold-expanded.[96]

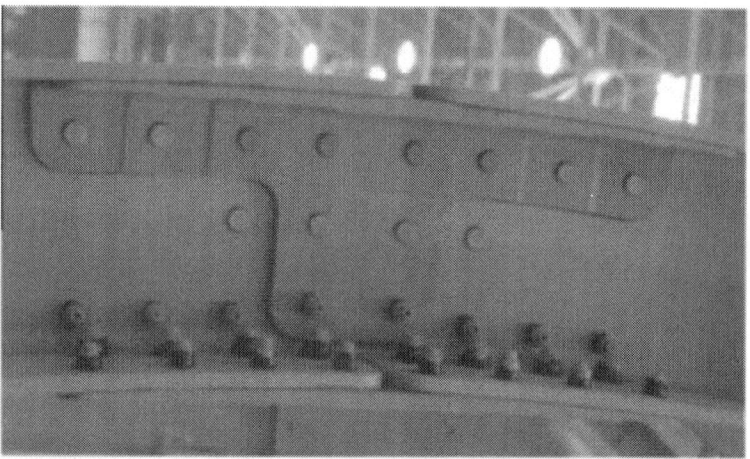

Figure 14: Multi-hole connection component in aircraft structures.

In early studies, the effect of the geometry of two adjacent cold expanded holes upon the resulting three-dimensional residual stress fields has been preliminarily examined considering both simultaneous and sequential cold expansion of the two holes.[15 and 22] These researches focus only on simulation of the residual stress around cold expanded holes. Amazingly, however, extensive search of the literatures indicates that no attempt has been made to experiment the cold expansion of numerous holes. Especially, the initiation and propagation mechanisms of fatigue cracks for the cold expanded adjacent holes need to be widely researched. It needs to do a lot of researches for improving the fatigue life of the adjacent holes part with cold expansion technology.

Sandwich Structure

The aircraft structure is composed of two or more parts connected together with assembly holes and fasteners. In order to improve the fatigue life of aircraft, it is essential for engineers to carry out anti-fatigue manufacture of assembly units. The current research of the cold expansion strengthening technology of holes is only limited to the single board, and it is not involved in the assembly structure. However, it is the key point to improve the fatigue life of the aircraft structure.

The cold expansion process for a multi-plate structure is required to complete expanding all holes of plates at one time, as shown in Fig. 15. Due to the differences of material properties and physical sandwich plates, different interferences should be used in the assembly units, especially for different materials. Consequently, the traditional cold expansion process of a single plate cannot be directly applied in sandwich plates. Experiments and simulation should be carried out to research the cold expansion technology of sandwich structures.

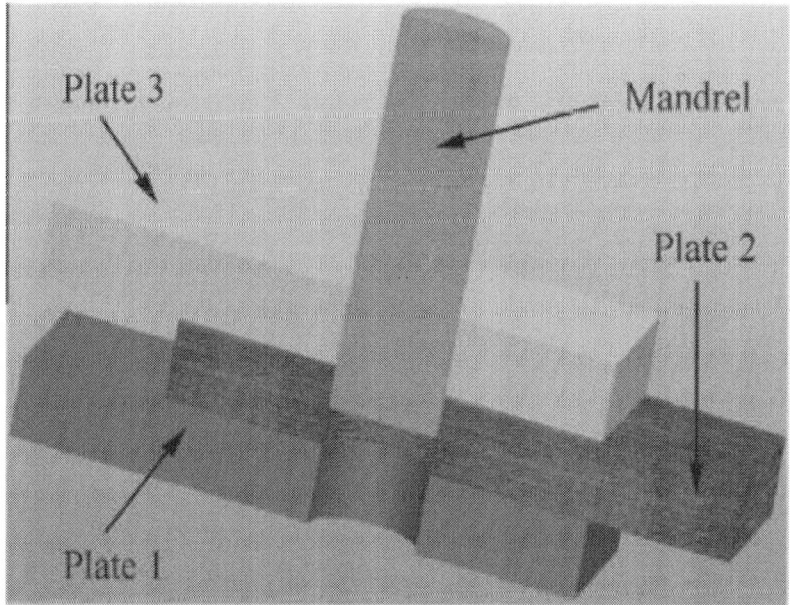

Figure 15: Cold expansion process for sandwich structure holes.

Effects of External Conditions

Cold expansion process has an obvious beneficial effect on the fatigue life of components with holes. However, the effect of cold expansion on the fatigue life is subjected by external conditions including bolt clamping, external stress, and working temperature. It should be pointed out that the researches of cold expansion strengthening technology have not been widely involved in the effects of external conditions on fatigue gain. In order to improve the fatigue life of a component under working conditions, it is necessary to carry out the initiation and propagation mechanisms of fatigue cracks and expansion process researches for connecting holes under certain environment.

Bolt Clamping

Among the most important connection structures in aircraft are mechanical joints, especially bolted joints. When a nut and a bolt are used to join mechanical members together, the nut is tightened by applying a torque, thus causing the bolt to axially stretch.[97] The result is that the bolt is left in tension and the mechanical members are compressed together. It is well known that the bolt clamping effect can decrease the stress concentration at the bolted hole region and thus increase the tensile and fatigue strengths of the joint.[98]

There are many researches about cold expansion and bolt clamping separately in the past decade. Nevertheless, studies about the combined effects of cold expansion and clamping torque in joints have not been carried out extensively. In fact, because a bolt is installed in a cold expanded hole, the compressive stress field of bolt clamping superimposes to the primary residual compressive stress field, and the result is that it can significantly affect the fatigue behavior of the cold expanded specimen. Moreover, the frictional contact between the washer of the bolt assembly and the specimen can bring further changes in the fatigue behavior.[99] Investigations of fatigue tests and numerical simulations should be conducted to study the combined effects of cold expansion and bolt clamping on fatigue life and failure mode.

External Stress

Fatigue failure is the most common failure mode of aircraft structural components. Fastener holes as natural stress concentrators are a basic factor for the fatigue strengths of these components.[45] Research shows that the cold expansion process always improve fatigue life compared to that of ''as drilled'' holes in all stress levels.[62 and 78] In a cold expansion process, residual stresses are those stresses existing in the structural components without an applied external load, and they form a self-equilibrated system of internal forces.[100] However, the external cyclic stress is the main factor affecting the fatigue lives of components.[101] Residual stress relaxation occurs around cold expanded holes under cyclic tension. That is to say, the beneficial anti-fatigue effect from the cold expansion process is influenced by external cyclic stress.[102] In spite of the extensive researches of the cold expansion process, the effect of stress levels on fatigue life after cold expansion has not been researched extensively. To efficiently improve fatigue life, the optimization of process parameters should be studied according to the load spectrum of aircraft in service.

Working Temperature

High temperatures can be generated near engine components as well as in the structure of an aircraft sitting idle on blacktop runways in the sun, and additionally, frictional heating may occur during supersonic flight.[103] When the speed of a supersonic aircraft reaches 2.4 Mach, temperatures arising from aerodynamic heating are anticipated to be 160 °C and higher.[104] Residual stresses are generated by the cold expansion process around expanded holes. However, there is a concern that the residual stresses may be subjected to relaxation under elevated temperature conditions and can therefore significantly affect the fatigue life.

One of the most important reasons for residual stress relaxation is metal material creep in service. The relaxation is caused not by external reasons, but by the natural property of non-ferrous alloys to change their internal equilibrium at room temperature.[105] The creep behavior of aluminium alloys is typical for high temperatures and its effect on residual stress relaxation is studied.106· 107 and 108 In current studies, cold expansion technology is employed usually in components which are exposed only to service conditions at room temperature. With regard to the problem that temperature fluctuations affect the fatigue resistance of a cold expanded hole, only a few papers have been published. Consequently, it is important to research whether fatigue life improvement from cold expansion is still retained at elevated temperatures.

CONCLUSIONS

(1) Cold expansion has been one key anti-fatigue manufacturing technology for light-weight, high-strength, and durable components in aviation industries. The fatigue lives of 7B50-T7451, 7050 aluminium alloy, and Ti6Al4V can be increased 28, 6, and 2.8 times, respectively, using hole cold expansion processes. These approaches including hole edge expansion, direct mandrel expansion, ball

expansion, and split sleeve expansion have been introduced and analyzed respectively. New cold expansion approaches have been put forward, through the analysis of the disadvantages existing in the current processes.

(2) Experimental investigations and finite element simulations have been made for better understanding of the anti-fatigue behavior with cold expansion technology. Residual stress and fatigue cracks are the key areas of current cold expansion research. On the basis of an analysis of the factors affecting the residual stress and the initiation and propagation of fatigue cracks, the current urgent demands for development of cold expansion have been described. Thus, studies of cold expansion technology should be carried out extensively according to structural characteristics and working conditions.

(3) With the development of aviation manufacturing technology, numerous new materials are widely utilized for commercial and military aircraft, including Al-Li alloys, titanium, composites, etc. Because of the differences of material properties, the traditional methods and process parameters of hole cold expansion are not applicable for them. Therefore, referencing current strengthening methods, novel methods and tools of hole cold expansion for new materials are designed necessarily according to the different material properties. The strengthening mechanism, parameter optimization, and effect of cold expansion need to be further studied and revealed through a large number of simulations and experimental researches.

ACKNOWLEDGEMENTS

The authors would like to thank the financial support of the Fund of National Engineering and Research Center for Commercial Aircraft Manufacturing of China (No. SAMC12-JS-15-021), the Funding of Jiangsu Innovation Program for Graduate Education of China(No. CXLX12_0137), and the Fundamental Research Funds for the Central Universities of China.

REFERENCES

1. Huang W, Wang TJ, Garbatov Y, Guedes Soares C. Fatigue reliability assessment of riveted lap joint of aircraft structures. Int J Fatigue 2012;43:54–61.
2. Yan CL, Liu KG. Theory of economic life Prediction and reliability assessment of aircraft structures. Chin J Aeronaut 2011;24(2):164–70.

3. Skorupa A, Skorupa M, Machniewicz T, Korbel A. Fatigue crack location and fatigue life for riveted lap joints in aircraft fuselage. Int J Fatigue 2014;58:209–17.

4. Chakherlou TN, Abazadeh B, Vogwell J. The effect of bolt clamping force on the fracture strength and the stress intensity factor of a plate containing a fastener hole with edge cracks. Eng Fail Anal 2009;16(1):242–53.

5. Liu J, Xu HL, Zhai HB, Yue ZF. Effect of detail design on fatigue performance of fastener hole. Mater Des 2010;31(2):976–80.

6. Liu J, Yue ZF, Liu YS. Surface finish of open holes on fatigue life. Theor Appl Fract Mech 2007;47(1):35–45.

7. Abazadeh B, Chakherlou TN, Alderliesten RC. Effect of interference fitting and/or bolt clamping on the fatigue behavior of Al alloy 2024–T3 double shear lap joints in different cyclic load ranges. Int J Mech Sci 2013;72:2–12.

8. Wang M. Principle & technology of anti-fatigue manufacture. 1st ed. Nanjing: Jiangsu Science and Technology Press; 1999, p. 296–303 [Chinese].

9. Jiang JF, Dong HY, Ke YL. Maximum interference fit size of hi-lock bolted joints. Chin J Mech Eng 2013;49(3):145–52 [Chinese].

10. Jiang JF, Dong HY, Bi YB. Analysis of process parameters influencing protuberance during interference fit installation of hilock bolts. Acta Aeronaut Astronaut Sin 2013;34(4):936–45 [Chinese].

11. Sara B, Ramin G, Mario G. Numerical and experimental analysis of surface roughness generated by shot peening. Appl Surf Sci 2012;258(18):6831–40.

12. Ren XD, Zhan QB, Yang HM, Dai FZ, Cui CY, Sun GF, et al. The effects of residual stress on fatigue behavior and crack propagation from laser shock processing-worked hole. Mater Des 2013;44:149–54.

13. Zhang XQ, Chen LS, Yu XL, Zuo LS, Zhou Y. Effect of laser shock processing on fatigue life of fastener hole. Trans Nonferrous Met Soc China 2014;24(4):969–74.

14. Barter S, Dixon B. Investigation using quantitative fractography of an unexpected failure in an F/A-18 centre fuselage bulkhead in the FINAL teardown program. Eng Fail Anal 2009;16(3):833–48.

15. Papanikos P, Meguid SA. Three-dimensional finite element analysis of cold expansion of adjacent holes. Int J Mech Sci 1998;40(10):1019–28.

16. Tolga D, Costas S. Recent developments in advanced aircraft aluminium alloys. Mater Des 2014;56:862–71.

17. Yang XY, Dong LW, Zheng XM, Zhao FX. Simulation of extrusion strengthening of pressure equalizing hole of an engine compressor disc and design of extrusion heads. J Aerosp Power 2013;28(8):1769–76 Chinese.

18. Liu J, Gou WX, Liu W, Yue ZF. Effect of hammer peening on fatigue life of aluminum alloy 2A12-T4. Mater Des 2009;30(6):1944–9.

19. Chakherlou TN, Aghdam AB. An experimental investigation on the effect of short time exposure to elevated temperature on fatigue life of cold expanded fastener holes. Mater Des 2008;29:1504–11.

20. Chakherlou TN, Aghdam AB, Akbari A, Saeedi K. Analysis of cold expanded fastener holes subjected to short time creep: finite element modelling and fatigue tests. Mater Des 2010;31(6):2858–66.

21. Liu J, Shao XJ, Liu YS, Yue ZF. Effect of cold expansion on fatigue performance of open holes. Mater Sci Eng A 2008;477 (1–2):271–6.

22. Kim C, Kim DJ, Seok CS, Yang WH. Finite element analysis of the residual stress by cold expansion method under the influence of adjacent holes. J Mater Process Technol 2004;153–154:986–91.

23. Maximov JT, Duncheva GV, Ganev N. Enhancement of fatigue life of net section in fitted bolt connections. J Constr Steel Res 2012;74:37–48.

24. Maximov JT, Duncheva GV. A new 3D finite element model of the spherical mandrelling process. Finite Elem Anal Des 2008;44(6–7):372–82.

25. Semari Z, Aid A, Benhamena A. Effect of residual stresses induced by cold expansion on the crack growth in 6082 aluminum alloy. Eng Fract Mech 2013;99:159–68.

26. Su M, Amrouche A, Mesmacque G, Benseddiq N. Numerical study of double cold expansion of the hole at crack tip and the influence on the residual stresses field. Comput Mater Sci 2008;41(3):350–5.

27. Leon A. Benefits of split mandrel coldworking. Int J Fatigue 1998;20(1):1–10.

28. Pavier MJ, Poussardand CGC, Smith DJ. Finite element simulation of the cold working process for fastener holes. J Strain Anal Eng Des 1997;32(4):287–300.

29. Karabina ME, Barlat F, Schultz RW. Numerical and experimental study of the cold expansion process in 7085 plate using a modified split sleeve. J Mater Process Technol 2007;189 (1–3):45–57.

30. O¨ zdemir AT, Hermann R. Effect of expansion technique and plate thickness on near-hole residual stresses and fatigue life of cold expanded holes. J Mater Sci 1999;34(6):1243–52.

31. Gopalakrishna HD, Murthy Narasimha HN, Krishna M, Vinod AV, Suresh AV. Cold expansion of holes and resulting fatigue life enhancement and residual stresses in Al 2024 T3 alloy – An experimental study. Eng Fail Anal 2010;17(2):361–8.

32. Zhang YW, Yang QX. An adaptive study of deformation theory in elastic–plastic computation for the process of coldworking fastener hole. Acta Aeronaut Astronaut Sin 1994;15(2):219–21 [Chinese].

33. Guo WL. Elastoplastic analysis of a finite sheet with an interference-fitted and coldworked hole. Acta Mech Sol Sin 1992;13(2):164–8 [Chinese].

34. Liu XL, Gao YK, Liu YT, Chen DF. 3D finite element simulation and experimental test on residual stress field by hole cold expansion. J Aeronaut Mater 2011;31(2):24–7 [Chinese].

35. Priest M, Poussard CG, Pavier MJ, Smith DJ. An assessment of residual-stress measurements around cold-worked Holes. Exp Mech 1995;35(4):361–6.

36. Pina JCP, Dias AM, de Matos RFP, Moreira PMGP, de Castro PMST. Residual stress analysis near a cold expanded hole in a textured alclad sheet using X-ray diffraction. Exp Mech 2005;45(1):83–8.

37. Tang C, Yu WM, Yun DZ. A method of technological elastoplastic and residual strain analyses of pre-worked holes. J Dalian Univ Technol 1998;38(1):25–8 [Chinese].

38. Xing WZ, Wu FM. Measurements and calculations of residual stresses at the edges of cold-worked holes. Acta Aeronaut Astronaut Sin 1991;12(2):B56–60 [Chinese].

39. Lin Y, Wu FM, Xing WZ. Experimental investigation on change of coldworking residual strain of hole in aluminium alloy sheet. Acta Aeronaut Astronaut Sin 1994;15(2):219–21 [Chinese].

40. Chen M, Fu SW, Zuo DW, Wang M. Applying photo-plastics in analyzing the mechanism of strengthen anti-fatigue in the case of cold-worked hole. J Shanghai Jiaotong Univ 2000;34(3):295–8 [Chinese].

41. Chen M, Fu SW, Zuo DW, Wang M. Mechanical analysis of cold-worked hole by photo-plastic. J Nanjing Univ Sci Technol 2000;24(2):121–5 [Chinese].
42. Yan CA. Photoelastic coatings technique and its application in engineering. 1st ed. Beijing: National Defense Industry Press; 2003. p. 183–198 [Chinese].

43. Hermann R. Three dimensional stress distributions around cold expanded holes in Aluminum alloys. Eng Fract Mech 1994;48(6):819–35.

44. de Matos PFP, Moreira PMGP, Camanho PP, de Castro PMST. Numerical simulation of cold working of rivet holes. Finite Elem Anal Des 2005;41:989–1007.

45. Kang JD, Johnson WS, Clark DA. Three-dimensional finite element analysis of the cold expansion of fastener holes in two aluminum alloys. J Eng Mater Technol 2002;124(2):140–5.

46. Mahendra Babu NC, Jagadish T, Ramachandra K, Sridhara SN. A simplified 3-D finite element simulation of cold expansion of a circular hole to capture through thickness variation of residual stresses. Eng Fail Anal 2008;15(4):339–48.

47. Maximov JT, Duncheva GV, Ganev N, Bakalova TN. The benefit from an adequate finite element simulation of the cold hole expansion process. Eng Fail Anal 2009;16(1):503–11.

48. Toktas I, O¨ zdemir AT. Artificial neural networks solution to display residual hoop stress field encircling a split-sleeve cold expanded aircraft fastener hole. Expert Syst Appl 2011;38(1):553–63.

49. Farhangdoost K, Hosseini A. The effect of mandrel speed upon the residual stress distribution around cold expanded hole. Proced Eng 2011;10:2184–9.

50. Duan MM, Geng XL, Zhang YJ, Huang JX, Yue ZF. Experimental and simulation study on evolution of stress and strain during the hole expansion. Eng Mech 2013;30(11):55–60 [Chinese].

51. Amrouchea A, Su M, Aid A, Mesmacque G. Numerical study of the optimum degree of cold expansion: application for the precracked specimen with the expanded hole at the crack tip. J Mater Process Technol 2008;197(1–3):250–4.

52. Zhao CM, Hu HY, Zhou YF, Gao YK, Ren XJ, Yang QX. Experimental and numerical investigation of residual stresses around cold extrusion hole of ultrahigh strength steel. Mater Des 2013;50:78–84.

53. Nigrelli V, Pasta S. Finite-element simulation of residual stress induced by split-sleeve cold-expansion process of holes. J Mater Process Technol 2008;205(1–3):290–6.

54. Ayatollahi MR, Arian Nik M. Edge distance effects on residual stress distribution around a cold expanded hole in Al 2024 alloy. Comput Mater Sci 2009;45(4):1134–41.

55. Maximov JT, Kuzmanov TV, Anchev AP, Ichkova MD. A finite element simulation of the spherical mandrelling process of holes with cracks. J Mater Process Technol 2006;171(3):459–66. 56. Amrouche A, Mesmacque G, Garcia S. Cold expansion effect on the initiation and the propagation of the fatigue crack. Int J Fatigue 2003;25(9–11):949–54.

57. Liu J, Wu HG, Yang JJ, Yue ZF. Effect of edge distance ratio on residual stresses induced by cold expansion and fatigue life of TC4 plates. Eng Fract Mech 2013;109:130–7.

58. Liu XL, Zhang Z, Tao CH. Fatigue fractography quantitative analysis. 1st ed. Beijing: National Defense Industry Press; 2010. p. 14 [Chinese].

59. de Castro PMST, de Matos PFP, Moreira PMGP. An overview on fatigue analysis of aeronautical structural details: open hole, single rivet lap-joint, and lap-joint panel. Mater Sci Eng A 2007;468–470:144–57.

60. Chakherlou TN, Vogwell J. The effect of cold expansion on improving the fatigue life of fastener holes. Eng Fail Anal 2003;10(1):13–24.

61. Ding CF, Zhao ZY, Song DY. Effect of coldworked holes on the initiation life and propagation life of fatigue cracks in two ultrahigh strength steels. Acta Aeronaut Astronaut Sin 1994;15(8):960–7 [Chinese].

62. Yan WZ, Wang XS, Gao HS. Effect of split sleeve cold expansion on cracking behaviors of titanium alloy TC4 holes. Eng Fract Mech 2012;88:79–89.

63. Chakherlou TN, Taghizadeh H, Aghdam AB. Experimental and numerical comparison of cold expansion and interference fit methods in improving fatigue life of holed plate in double shear lap joints. Aerosp Sci Technol 2013;29(1):351–62.

64. Viveros KC, Ambriz RR, Amrouche A, Talha A, Garcı´ad C, Jaramillo D. Cold hole expansion effect on the fatigue crack growth in welds of a 6061–T6 aluminum alloy. J Mater Process Technol 2014;214(11):2606–16.

65. Liu YS, Shao XJ, Liu J, Yue ZF. Finite element method and experimental investigation on the residual stress fields and fatigue performance of cold expansion hole. Mater Des 2010;31(3):1208–15.

66. Yanishevsky M, Li G, Shi GQ, Backman D. Fractographic examination of coupons representing aircraft structural joints with and without hole cold expansion. Eng Fail Anal 2013;30:74–90.

67. de Matos PFP, McEvily AJ, Moreira PMGP, de Castro PMST. Analysis of the effect of cold-working of rivet holes on the fatigue life of an aluminum alloy. Int J Fatigue 2007;29(3):575–86.

68. de Matos PFP, Moreira PMGP, Pina JCP, Dias AM, de Castro PMST. Residual stress effect on fatigue striation spacing in a cold-worked rivet hole. Theor Appl Fract Mech 2004;42(2):139–48.

69. Daniel HS, Michael RH, James CNJ. Correlation of onedimensional fatigue crack growth at cold-expanded holes using linear fracture mechanics and superposition. Eng Fract Mech 2011;78(7):1389–406.

70. Lacarac VD, Smith DJ, Pavier MJ, Priest M. Fatigue crack growth from plain and cold expanded holes in aluminium alloys. Int J Fatigue 2000;22(3):189–203.

71. Chen L, Cai LX, Yao D. A new method to predict fatigue crack growth rate of materials based on average cyclic plasticity strain damage accumulation. Chin J Aeronaut 2013;26(1):130–5.

72. Wang LQ, Gai BZ. Numerical computation of stress intensity factors for bolthole corner crack in mechanical joints. Chin J Aeronaut 2008;21(5):411–6.

73. Pasta S. Fatigue crack propagation from a cold-worked hole. Eng Fract Mech 2007;74(9):1525–38.

74. Newman Jr JC, Daniewicz SR. Predicting crack growth in specimens with overloads and cold-worked holes with residual stresses. Eng Fract Mech 2014;127:252–66.

75. LaRue JE, Daniewicz SR. Predicting the effect of residual stress on fatigue crack growth. Int J Fatigue 2007;29(3):508–15.

76. Fan J, Li FG, Li J, Wang SG. Study of local cold working and tensile test for 7050 high strength aluminum alloy hole plate. Rare Met Mater Eng 2012;41(6):978–82 [Chinese].

77. Gong P, Zheng LB, Zhang K, Yi LN, Song DY. Effects of hole cold-expansion on microstructure and fatigue property of 7B50- T7451 aluminium alloy plate. J Aeronaut Mater 2011;31(4):45–50 [Chinese].

78. Burlata M, Julien D, Le´vesque M, Bui-Quoc T, Bemard M. Effect of local cold working on the fatigue life of 7475–T7351 aluminium alloy hole specimens. Eng Fract Mech 2008;75(8):2042–61.

79. Wu H. On the prediction of initiation life for fatigue crack emanating from small cold expanded holes. J Mater Process Technol 2012;212(9):1819–24.

80. Ling C, Zhang BF, Zheng XL. Method for predicting fatigue crack initiation life of elements subjected to CEH under variableamplitude loading. Acta Aeronaut Astronaut Sin 1991;27(1):A75–7 [Chinese].

81. Shao YS, Yang QX. A simple study on fatigue life for coldworked fastener holes. Acta Aeronaut Astronaut Sin 1990;11(12):B602–5 [Chinese].

82. Guan ZD, He QZ, Li ZN. A reliability analysis method on the fatigue lives of interference-fit or coldworking strengthened parts. Acta Aeronaut Astronaut Sin 1994;15(3):344–7 [Chinese].

83. Boni L, Lanciotti A, Polese C. Some contraindications of hole expansion in riveted joints. Eng Fail Anal 2014;46:140–56.

84. Giglio M, Lodi M. Optimization of a cold-working process for increasing fatigue life. Int J Fatigue 2009;31(11–12):1978–95.

85. Maximov JT, Duncheva GV, Amudjev IM. A novel method and tool which enhance the fatigue life of structural components with fastener holes. Eng Fail Anal 2013;31:132–43. 86. Maximov JT, Duncheva GV, Ganev N, Amudjev IM. Modeling of residual stress distribution around fastener holes in thin plates after symmetric cold expansion. J Braz Soc Mech Sci Eng 2014;36(2):355–69.

87. Cao CX. One generation of material technology, one generation of large aircraft. Acta Aeronaut Astronaut Sin 2008;29(3):701–6 [Chinese].

88. Xu JH, Zhang ZW, Fu YC. Review and prospect on high efficiency profile grinding of nickel-based superalloy. Acta Aeronaut Astronaut Sin 2014;35(2):351–60 [Chinese].

89. Song DY, Luo ZP, Yang YR, Zhang SQ. Microstructure of the hole expansion strengthened layer of high temperature alloy GH169. Acta Aeronaut Astronaut Sin 1996;17(1):125–8 [Chinese].

90. Tian F, Yang H. Effects of hole-expansion fatigue property of AM50 magnesium alloy. Surf Technol 2014;43(1):55–8 [Chinese].

91. Sun ZQ, Huang MH. Fatigue crack propagation of new aluminum lithium alloy bonded with titanium alloy strap. Chin J Aeronaut 2013;26(3):601–5.

92. Wei JC, Jiao GQ, Jia PR, Huang T. The effect of interference fit size on the fatigue life of bolted joints in composite laminates. Compos B Eng 2013;53:62–8.

93. Liu LQ, Zhang JQ, Chen KK, Wang H. Combined and interactive effects of interference fit and preloads on composite joints. Chin J Aeronaut 2014;27(3):716–29.

94. Faraz A, Biermann D, Weinert K. Cutting edge rounding: An innovative tool wear criterion in drilling CFRP composite laminates. Int J Mach Tools Manuf 2009;49(15):1185–96.

95. Wang YB, Zhang QC. Composite structures connection technology. 1st ed. Beijing: National Defense Industry Press; 1992. p. 121–122 [Chinese].

96. Papanikos P, Meguid SA. Elasto-plastic finite-element analysis of the cold expansion of adjacent fastener holes. J Mater Process Technol 1999;92–93:424–8.

97. Chakherlou TN, Mirzajanzadeh M, Vogwell J, Abazadeh B. Investigation of the fatigue life and crack growth in torque tightened bolted joints. Aerosp Sci Technol 2011;15(4):304–13.

98. Chakherlou TN, Alvandi-Tabrizi Y, Kiani A. On the fatigue behavior of cold expanded fastener holes subjected to bolt tightening. Int J Fatigue 2011;33(6):800–10.

99. Chakherlou TN, Shakouri M, Akbari A, Aghdam AB. Effect of cold expansion and bolt clamping on fretting fatigue behavior of Al 2024–T3 in double shear lap joints. Eng Fail Anal 2012;25:29–41.

100. Maximov JT, Duncheva GV, Mitev IN. Modelling of residual stress relaxation around cold expanded holes in carbon steel. J Constr Steel Res 2009;65(4):909–17.

101. Warner JJ, Clark PN, Hoeppner DW. Cold expansion effects on cracked fastener holes under constant amplitude and spectrum loading in the 2024–T351 aluminum alloy. Int J Fatigue 2014;68:209–16.

102. Chakherlou TN, Shakouri M, Aghdam AB, Akbari A. Effect of cold expansion on the fatigue life of Al 2024–T3 in double shear lap joints: experimental and numerical investigations. Mater Des 2012;33:185–96.

103. Clark DA, Johnson WS. Temperature effects on fatigue performance of cold expanded holes in 7050–T7451 aluminum alloy. Int J Fatigue 2003;25(2):159–65.

104. Lacarac VD, Smith DJ, Pavier MJ. The effect of cold expansion on fatigue crack growth from open holes at room and high temperature. Int J Fatigue 2001;23:161–70.

105. Maximov JT, Duncheva GV, Anchev AP. An approach to modeling time-dependent creep and residual stress relaxation around cold worked holes in aluminium alloys at room temperature. Eng Fail Anal 2014;45:1–14.

106. Minguez JM, Vogwell J. Fatigue life of an aerospace aluminium alloy subjected to cold expansion and a cyclic temperature regime. Eng Fail Anal 2006;13(6):997–1004.

107. Aghdam AB, Chakherlou TN, Saeedi K. An FE analysis for assessing the effect of short-term exposure to elevated temperature on residual stresses around cold expanded fastener holes in aluminum alloy 7075–T6. Mater Des 2010;31(1):500–7.
108. Maximov JT, Duncheva GV, Anchev AP, Ichkova MD. Modeling of strain hardening and creep behaviour of 2024T3 aluminium alloy at room and high temperatures. Comput Mater Sci 2014;83:381–93.

CITATION

Fu Yucan , Ge Ende, Su Honghua, Xu Jiuhua, Li Renzheng, cold expansion technology of connection holes in aircraft structures: A review and prospect, doi:10.1016/j.cja.2015.05.006

CHAPTER 3

Aeromtp: A Fountain Code-Based Multipath Transport Protocol for Airborne Networks

Li Jie, Gong Frling, Sun Zhiqiang, Liu Wei, Xie Hongwei

College of Mechatronic Engineering and Automation, National University of Defense Technology, Changsha 410073, China

ABSTRACT

Airborne networks (ANs) are special types of ad hoc networks that can be used to enhance situational awareness, flight coordination and flight efficiency in civil and military aviation. Compared to ground networks, ANs have some unique attributes including high node mobility, frequent topology changes, mechanical and aerodynamic constrains, strict safety requirements and harsh communication environment. Thus, the performance of conventional transmission control protocol (TCP) will be dramatically degraded in ANs. Aircraft commonly have two or more heterogeneous network interfaces which offer an opportunity to form multiple communication paths between any two nodes in ANs. To satisfy the communication requirements in ANs, we propose aeronautical multipath transport protocol (AeroMTP) for ANs, which effectively utilizes the available bandwidth and diversity provided by heterogeneous wireless paths. AeroMTP uses fountain codes as forward error correction (FEC) codes to recover from data loss and deploys a TCP-friendly rate-based congestion control mechanism for each path. Moreover, we design a packet allocation algorithm based on optimization to minimize the delivery time of blocks. The performance of AeroMTP is evaluated through OMNeT++ simulations under a variety of test scenarios. Simulations demonstrate that AeroMTP is of great potential to be applied to ANs.

INTRODUCTION

Airborne networks (ANs), also called aeronautical ad hoc networks (AANETs), are special types of mobile ad hoc networks (MANET) that can be used on air vehicles, moving platforms, and ground stations to enhance situational awareness, flight coordination, and flight efficiency.[1] Due to the self-organizing multi-hop feature, ANs can greatly expand the connectivity over the line-of-sight limitation of radio wave and the limitation of transmission range. This benefit has attracted the attention of researchers in civil aviation to apply them for aeronautical communication in remote area such as ocean, deserts, etc. Data communications through ANs can provide the internet connection and periodic downloads of "black box" data in real time,[2 and 3] which can avoid spending too much time and effort on the search for black boxes after air crash, such as the search for the black boxes of Air France Flight 447 and Indonesia AirAsia Flight 8501. In military fields, current military airborne communication systems provide only their own mission specific implementations, and provide limited interoperability and autonomous routing capability. ANs are envisioned as hierarchical networks with the capacity of supporting diverse heterogeneous networks operating with various protocols and communication links. Moreover, ANs can satisfy the stringent requirements of military networks and reduce sensor-to-shorter timeline by combining data from disparate sensors, air platforms, and ground stations.[4 and 5]

ANs are formed rapidly in the airspace and have some attributes including high node mobility, frequent topology changes, mechanical and aerodynamic constrains, strict safety requirements, and harsh communication environment. Compared with general wireless networks, the most important characteristics and challenges posed by ANs are list as follows:

(1) Long and volatile delays: aircraft in ANs connect to each other mainly through the MANET formed by airborne platforms via Radio Frequency (RF) or Optical links. Due to the highly dynamic topology of the MANET, the total hops between any two aircraft may change frequently, which results in a volatile end-to-end delay. In addition, the long communication distance between two aircrafts brings about a long propagation delay, especially when Satellite Communication (SATCOM) links are used.

(2) High link error rates: in ANs, the wireless channels are subject to harsh communication environment such as interference, jamming, and channel impairments, which results in a high link error rate.

(3) Frequent blackouts: due to the harsh communication environment and high mobility, airborne links are likely to experience intermittent link blockages and signal losses. Consequently, the end-to-end connectivity between any two nodes may be broken frequently.

(4) Bandwidth asymmetry: in ANs, ground-based transceivers commonly have large steerable dish antennas, but aircraft are usually equipped with small omnidirectional antennas, which brings asymmetric links into ANs.

In order to provide interconnectivity with space and terrestrial networks, the future ANs will be based on the Internet protocol (IP).[2 and 5] An efficient and fair transport protocol should be designed for ANs to support Internet and data services. However, conventional transmission control protocol (TCP) would see variable capacity, unpredictable packet erasures and volatile delays from such communication paths in ANs. Frequent blackouts of the communication paths make TCP go to slow start phase frequently, leading to a severe reduction of the throughput. For an asymmetric link, TCP may cause reverse channel to be congested, resulting in packet losses in the reverse channel. Hence, ANs need a transport protocol that can tolerate volatile channel conditions and address the bandwidth asymmetry problem while maintaining stable goodput and latency for the networks. Airborne platforms commonly have two or more heterogeneous network interfaces (e.g., RF, Optical and SATCOM) which offer an opportunity to form multiple independent communication paths for any two nodes in ANs.[6] Each of those paths experiences blackouts, losses, delay and capacity variations independent of other paths. This type of network that can simultaneously establish multiple independent paths between any two hosts is called multihoming network. Studies have shown that transmissions over multiple paths can significantly improve the transmission capacity and quality.[7] Therefore, ANs can operate as multihoming networks to improve the performance, and then transport protocols for ANs should counter the volatility of a single path by transmitting data across different paths intelligently and leverage diversity among paths to yield stable and high goodput.

In this paper, a fountain code-based multipath transport protocol is proposed for ANs, called aeronautical multipath transport protocol (AeroMTP). AeroMTP uses fountain code as the packet-level forward error correction (FEC) code and deploys a TCP-friendly rate-based congestion control mechanism for each path. The congestion control mechanism has the ability to control the amount of traffic on the reverse channel of a path, which well solves the bandwidth asymmetry problem in ANs. A novel packet allocation algorithm is designed for AeroMTP based on optimization to minimize the delivery time of blocks. The rest of the paper is organized as follows. We discuss the related work in Section 2 and present the architecture of AeroMTP in Section 3. The overall design of AeroMTP is addressed in Section 4. We present and discuss our simulation study in Section 5 and conclude the work in Section 6.

RELATED WORK

Recently, some domain-specific protocols are proposed for ANs.[8 and 9] Rohrer et al. presented a transport protocol, AeroTP, for the aeronautical telemetry network.[9] AeroTP is designed to meet the needs of the highly-dynamic network environment while being TCP-friendly to allow seamless splicing with conventional TCP at the network edge. AeroTP uses a strong cyclic redundancy check (CRC) to detect bit errors in the wireless channel, which allows a corrupted payload to be corrected on an end-to-end basis using FEC. However, the payload CRC can only protect the integrity of the data, but cannot guard against packet losses. Consequently, the performance of AeroTP degrades significantly due to frequent retransmissions of lost packets in the network with high loss rate. Furthermore, since AeroTP transmits data via a single path, the performance is also seriously affected by the volatile path conditions in the harsh aeronautical environment. Many efforts have been made to explore the use of multiple paths simultaneously to improve the performance of networks. An existing multipath transport protocol that supports multistreaming and multihoming is stream control transmission protocol (SCTP).[10] However, in the basic SCTP design, other paths are only considered as the backups of a primary path, and concurrent multipath transfer (CMT) is not supported. CMT-SCTP[11] extends SCTP by adding CMT support, which is in the process of IETF standardization. Ford et al. attempted to design and implement a deployable multipath TCP that is backward-compatible with TCP which has been standardized in Ref.[12]. However, the above-mentioned protocols suffer from performance degradation due to frequent retransmissions and reordering in the network with high loss and delay. In order to improve the performance over lossy networks, packet-level FEC code is introduced into multipath protocols.[13, 14 and 15] In these protocols, application layer data is divided into blocks which are then converted to a series of encoded packets. As long as the receiver successfully receives a certain number of encoded packets for a block, this block can be recovered and transmitted to the application layer. Multi-path loss-tolerant (MPLOT) transport protocol[13] uses a fixed rate coding scheme, which does not have good performance when the path quality decreases sharply. As a rateless code, fountain code[16] is used for packet-level FEC in heterogeneous multipath transport protocol (HMTP)[14] and fountain code based multipath TCP (FMTCP),[15] which can work better than the fixed rate FEC code over lossy networks. However, the multipath transport protocols in Ref.[13, 14 and 15] are not suitable for ANs in the harsh communication environment. On the one hand, these protocols use TCP-like window-based congestion control for each path. As channel bandwidth, link delay and loss rate increase, the window-based congestion control shows inefficiency and instability.[17] The window-based congestion control used in Ref.13, 14 and 15 can also not counter some of the challenges posed by ANs such as frequent blackouts and bandwidth asymmetry. On the other hand, the packet allocation schemes used in these protocols are to deliver the encoded packets in a short time by sending them via "better" paths, which cannot guarantee that a block can be successfully delivered in a minimal amount of time over multiple volatile paths in ANs.

ARCHITECTURE OF AEROMTP

Fig. 1 illustrates a typical "vision" of ANs described in Ref.[2]. We can see that it mainly comprised of aircraft, satellites, ground stations, and high altitude platforms (HAPs). Satellite or HAP serves as a communication relay in ANs, which can provide stable data relay services for aircraft in the coverage areas. In the airspace, aircraft form a MANET via omnidirectional links (e.g., very high frequency (VHF) data links) or directional links (e.g., Ka-band and optical data links). The MANET can provide high bandwidth connections for aircraft, however, communication paths via the MANET may suffer from frequent breakouts and volatile delays due to the high mobility of aircraft. Consequently, multiple independent communication paths can be simultaneously established between any two nodes via the communication relays and MANET, which offers an opportunity to apply multipath transport protocol to ANs.

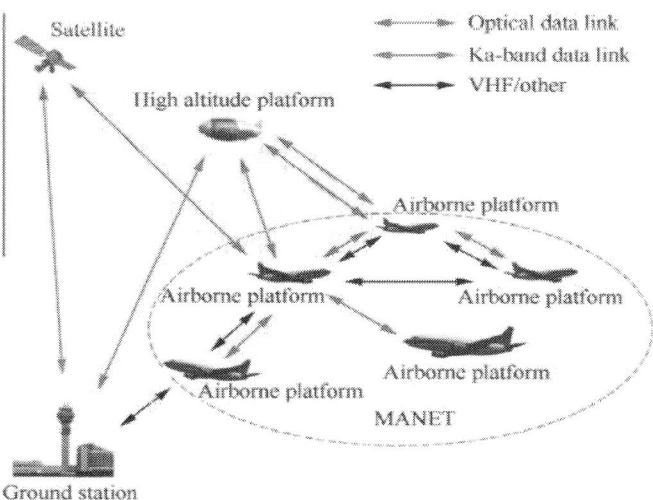

Figure 1: Airborne networks.

In the harsh aeronautical environment, a single communication path may suffer from challenges such as long/volatile delays, high link error rates, frequent blackouts and bandwidth asymmetry which can significantly degrade the performance. As a multipath transport protocol, AeroMTP can effectively solve these problems and improve the total goodput for ANs. The sender-side architecture of AeroMTP is illustrated in Fig. 2. The fountain code is introduced into AeroMTP to improve the transmission quality. Compared to a fixed-rate coding scheme, the rateless fountain code has the benefit of changing the coding rate on the fly based on the receiving quality while introducing very low overhead. A byte steam from applications is divided into blocks, which are taken as the input of the encoding module. After the encoding, each block is converted to a series of encoded symbols which are packetized as encoded packets. According to the quality of each path, the packet allocation module implements a packet mapper, which maps the encoded packets to different paths. The packets mapped to a path are then transmitted to the receiver via the individual path at a transmission rate determined by the rate control module. On the receiver side, when enough encoded packets for a block are successfully received, the receiver implements decoding process on those packets to recover the block. Once decoded successfully, the data can be transmitted to the application layer.

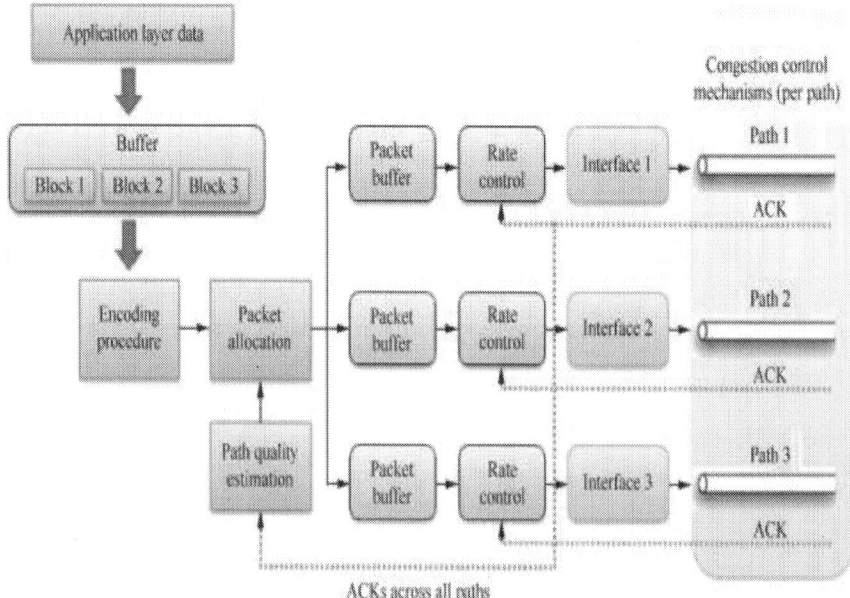

Figure 2:.Architecture of AeroMTP (sender side).

AEROMTP DESIGN

As shown in Fig. 2, the encoding procedure, packet allocation and congestion control mechanism for each path are the three key components in AeroMTP. The encoding procedure uses fountain codes as the FEC code to reduce the number of packet retransmissions in the lossy channel. The packet allocation component determines the total number of encoded packets generated for a block and allocates them to different paths to improve the total goodput. A rate-based additive-increase multiplicative-decrease (AIMD) congestion control mechanism with the capability to control the amount of traffic on the reverse channel is deployed for each path. Next, we describe how these components work in detail.

Coding Analysis

AeroMTP uses packet-level FEC coding to recover from data losses in channels with high loss rate. In this way, if some encoded packets for a block are lost, the receiver can still be able to recover this block when enough encoded packets are successfully received. FEC codes have two categories: fixed-rate and rateless. Let k be the block size, i.e., the block contains k symbols. For a fixed-rate code, the block is encoded into k_b encoded packets, where k/k_b is the coding rate, and the receiver can recover the block if any k of the k_b encoded packets are successfully received. The sender should retransmit some encoded packets for the block if the number of packets lost is larger than k_b-k. This coding scheme works well on channels whose loss rate does not change rapidly. Since the coding rate of the fixed-rate coding scheme cannot be altered during the transmission, the fixed-rate coding schemes are no longer suitable for ANs where the loss rate may vary over a large range. In rateless coding, arbitrary number of encoded packets can be generated for a block. Thus, the rateless coding has no fixed rate and its rate adapts between 1 and 0. With rateless codes, the coding rate can be adjusted according to the channel conditions to improve the goodput. As a kind of rateless codes, fountain code is used in AeroMTP. Fountain codes offer a very promising future for ameliorating existing data packet transmission techniques. [16] Fountain code has different implementations, such as Luby transform (LT) code [18] and Raptor code. [19] Since Raptor code extends LT code and has a better performance, we adopt the Raptor code for AeroMTP. Raptor code is a more practical fountain code which has been standardized in RFC5053 [20] and RFC6330. [21]

In Raptor code, a block containing k symbols can generate arbitrary number of encoded packets. The receiver can recover this block with a very high probability when $k' = k(1 + \varepsilon)$ encoded packets are received, where ε is the coding redundancy. In practice, a fountain code-based protocol using Raptor code can be implemented with $\varepsilon \approx 0.05$ [16] and we set $\varepsilon = 0.05$ in this paper. In lossy channels, some encoded packets would be lost and the sender should send more than k' encoded packets for a block to ensure that the receiver can successfully receive at least k' encoded packets. How to determine the total number of encoded packets sent for a block via lossy channels will be discussed in Section 4.3.

Design of Per-Path Congestion Control

Different paths in ANs may have different characteristics or temporarily experience different conditions (e.g., different bandwidths, end-to-end delay, blackouts, etc.). If the protocol responds congestion at the aggregate level, the performance level will be dominated by the worst path. Consequently, congestion control in AeroMTP is performed by each path, independent of the other paths.

Since the window-based congestion control is not suitable for the environments with high delay and link error rate[17], AeroMTP deploys a rate-based congestion control mechanism for each path. The rate-based mechanism directly controls the transmission rate of the path to avoid congestion, based on the feedback from the network. The congestion detection method is the core of a congestion control mechanism. TCP-like congestion control mechanisms depend on packet loss to respond to congestion, and are unable to distinguish between losses due to congestion and channel errors. Explicit congestion notification (ECN)[22] can be used to indicate incipient congestion, which allows us to isolate losses due to channel errors. However, there is always the possibility of losing ECN marked packets due to link errors, which leads to inaccurate congestion control. The methods that measure the observed rate at receiver side to determine the available bandwidth23 [and] 24 are also not suitable for ANs, because the end-to-end delay of paths in ANs may vary frequently and sharply, and the arrival time differences between consecutive packets could not express available bandwidth accurately. In Ref.[25], Akan et al. used NIL segments to detect congestion. The NIL segments are carried by IP packets, which are marked as high and low priority using TOS field in the IP packet header. Assume that all routers in the connection path have priority-queuing capability. The low-priority NIL segments are discarded first in case of a congestion. Low- and high-priority NIL segments are transmitted simultaneously with the same rate. The only reason for low- and high-priority NIL segments to have different packet loss rates is the additional loss experienced by low-priority segments due to congestion. This reasoning constitutes the basis for the congestion detection method via NIL segments transmission. However, the sender should send NIL segments periodically for the congestion detection, which may cause a decrease in the throughput of actual data packets.

Similar to the NIL segments, AeroMTP alternately marks the packets as high and low priority for each path (see Fig. 3), and the low-priority packets are discarded first in the case of a congestion. Since the encoded packets for a block are equally important for decoding the block, an encoded packet can be marked as high or low priority. There are two exceptions: the control packets (e.g., acknowledgment packets) and the additional encoded packets retransmitted for a unrecovered block are only marked as high priority. Thus, we can implement the congestion detection without any extra segments. The sender transmits packets with the rate r_i (in packets/s) via path i, i.e., the low- and high-priority packets are transmitted simultaneously with the same rate $r_i/2$. Each packet also needs a field in its header to indicate the instantaneous transmission rate (ITR) when it is transmitted to the receiver.

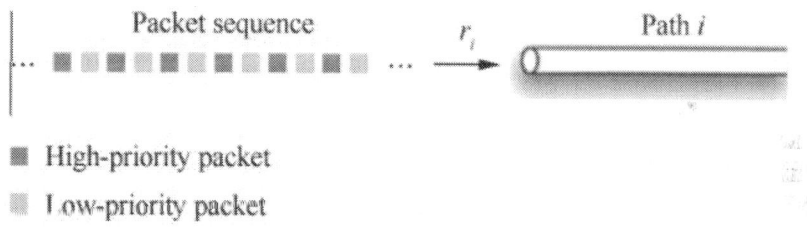

Figure 3: Packet sequence with low- and high-priority packets.

The receiver counts the number of received low (n_{low}) and high (n_{high}) priority packets from path i in a sliding time window of T_i, and calculates the average transmission rate of those packets, denoted by r_{ave}, according to their ITR values. The value of r_{ave} can be calculated as

$$r_{ave} = \frac{\sum_{j=1}^{n_{low}+n_{high}} ITR_j}{n_{low} + n_{high}}$$

(1)

where ITR_j represents ITR value of the jth packets received during the time T_i. Then, the receiver sends back an acknowledgment (ACK) with the 3-tuple (n_{low}, n_{high}, r_{ave}) every τ period. Let H be the ratio of received low- and high-priority packets, i.e., $H = n_{low}/n_{high}$. Since the low- and high-priority packets experience the same packet loss rate, the only reason for $n_{low} < n_{high}$ is the congestion along the path, i.e., we can infer that a congestion exists if $H < 1$. Thus, the sender can adjust transmission rate according to the value of H. In order to avoid unnecessary rate decreasing, decision is made via a comparison between H and the rate decrease threshold h_d which is a little smaller than 1.

In a shared network, all traffic flows are expected to be TCP friendly,[26] i.e., rate adjustment should result in a fair share of bandwidth for the TCP-based flows along the same path. In TCP, the congestion window is increased by one packet per round-trip time (RTT) if no congestion is experienced and reduced by half when congestion is detected. Let RTT_i be the value of RTT for path i. In order to match the behavior of AeroMTP with the behavior of TCP, the transmission rate r_i is increased by $1/RTT_i$ per RTT_i in a step-like fashion if no congestion is experienced along the path i. When the sender receives an ACK carrying $(n_{low}, n_{high}, r_{ave})$ from path i, where $n_{low}/n_{high} < h_d$, the sender infers that congestion is experienced along path i if the packets are transmitted at rate r_{ave}. Then, the transmission rate r_i is decreased multiplicatively, i.e., $r_i = r_{ave}/2$, to match with the behavior of TCP. In order to accurately update the transmission rate r_i, the ACK period τ_i for path i should be smaller than the sliding time window T_i, i.e., $\tau_i < T_i$. In AeroMTP, the value of RTT_i is estimated the same way as performed in TCP.

Based on the ACK with $(n_{low}, n_{high}, r_{ave})$, we can also estimate loss rates for each path. Assume that the sender receives an ACK carrying $(n_{low}, n_{high}, r_{ave})$ from path i. If $n_{low}/n_{high} > h_d$, we can infer that congestion is not experienced when the packets are transmitted at rate r_{ave} via path i. Then, $p_{i,inst}$, the instantaneous loss rate of path i, can be calculated as

$$p_{i,inst} = 1 - \frac{n_{low} + n_{high}}{r_{ave} T_i} \tag{2}$$

The running estimate of loss rates, p_i, is obtained by smoothing using an exponential weighted moving average (EWMA) method with a parameter value of 0.5.

If the sender does not receive any ACKs within T_i period, it infers this condition as blackout. During blackout, the sender stop sending any packets (set $r_i = 0$). The receiver also infers blackout after not receiving any packets from the sender for period of time T_i. Then, it starts to transmit ACKs with $(0, 0, 0)$ periodically, which are called Zero ACKs. The objective of Zero ACKs is to help the sender to capture accurate information regarding the blackout situation and act accordingly. After the duration of the blackout, the sender starts to receive Zero ACKs which indicates that the blackout is over. Then, the sender sets r_i to be the average transmission rate before the blackout and starts to transmit packets.

As shown in Fig. 4, the rate-based congestion control in AeroMTP can be modeled as a finite state machine model with three states: INITIAL, STEADY and BLACKOUT. At the sender side, the rate-based congestion control algorithm is summarized in Algorithm 1 as shown in Table 1. Let t represent the current system time and t_0 be the initial time instant. Let b_i (in packets/s) be the average transmission rate for path i which is calculated from the instantaneous transmission rate r_i using EWMA with a parameter value of 0.5. In the beginning of a new connection via path i, the sender must set the initial transmission rate value r_{ini} for r_i. The choice of the initial value r_{ini} is important, because if the value is higher than the available bandwidth, the new connection will cause network congestion; otherwise, resource utilization will be very low and will stay low for a time interval proportion to the bandwidth-delay product. Therefore, the initial value r_{ini} is set to the value that can meet the minimum requirements of the application. Then, the sender goes to STEADY state where r_i is decreased multiplicatively or increased additively according to the value of H every RTT_i period. If the sender does not receive any ACKs for a certain period of time T_i, it goes to BLACKOUT state and stops sending packets (set $r_i = 0$ and $b_i = 0$). Let b_{pre} (in packets/s) be the average transmission rate before the BLACKOUT. If some Zero ACKs are received, the sender sets r_i and b_i as the value of b_{pre}, and goes back to STEADY state.

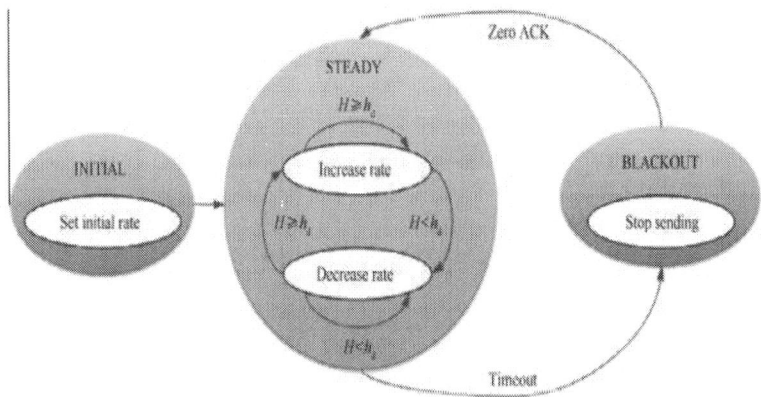

Figure 4: Operation state diagram of the rate-based congestion control mechanism.

Table 1: Rate-based congestion control for path i on sender side.

Algorithm 1 rate-based congestion control for path i on sender side
1. set $r_i = r_{ini}$
2. set $b_i = r_{ini}$
3. set $t_{next\_increase} = t_0 + RTT_i$
4. set state = STEADY
5. **while** (1) **do**
6. **if** state = = STEADY **then**
7.   **if** receive an ACK with $H < h_d$**then**
8.   set $r_i = r_{ave}/2$
9.   set $b_i = 0.5b_i + 0.5r_i$
10.   set $t_{next\_increase} = t + RTT_i$
11.   **else if** $t \geqslant t_{next\_increase}$
12.   **if** no ACK period is longer than T_i**then**
13.   set state = BLACKOUT
14.   set $r_i = 0$
15.   set $b_{pre} = b_i$
16.   set $b_i = 0$
17.   **else**
18.   set $r_i = r_i + 1/RTT_i$.
19.   set $b_i = 0.5b_i + 0.5r_i$.
20.   set $t_{next\_increase} = t + RTT_i$.
21.   **end if**
22.   **end if**
23. **else**
24.   **if** Zero ACKs are received **then**
25.   set state = STEDAY
26.   set $r_i = b_{pre}$.
27.   set $b_i = b_{pre}$.
28.   set $t_{next\_increase} = t + RTT_i$.
29.   **end if**
30. **end if**
31. **end while**

There commonly exist asymmetric links in ANs and AeroMTP also needs a mechanism to address the bandwidth asymmetry problem. We assume that encoded packets and ACKs have the same packet size. Considering that the ACK period of a flow via path i is τ_i, the ratio of the traffic in the forward and revise channel for path i is $C_i\tau_i$: n_{flow}, where $C_i$denotes the capacity of the forward channel for path i and n_{flow} represents the number of flows transmitted through path i. If the bandwidth asymmetry of path i with n_{flow} flows   is   higher   than $C_i\tau_i$: n_{flow},   sending   ACKs every τ_i period can cause reverse channel of path $i$to be congested, resulting in packet losses in the reverse link. In order to avoid this problem, we introduce an integer d as the delaying factor to adjust the ACK period. As shown in Algorithm 1 (Table 1), if no congestion is experienced along path i, the receiver can send ACKs every $d\tau_i$ period, which reduces the traffic in the revise channel and does not influence the rate-based   congestion   control   mechanism.   When   congestion   is experienced, ACKs are still sent every τ_i period. Moreover, in order to avoid disturbing the blackout detecting, the delaying factor d should satisfy $d\tau_i < T_i$. The ACK sending algorithm for path i is shown in Algorithm 2 (Table 2). Therefore, the traffic on the reverse channel can be controlled by adjusting the delaying factor d, which well solves the bandwidth asymmetry problem in ANs.

Table 2: ACK sending scheme for path i on receiver side.

Algorithm 2 ACK sending scheme for path i on receiver side
1. set $n = 0$
2. set $t_{next\_ack} = t_0 + \tau_i$
3. **while** (1) **do**
4. if$t \geqslant t_{next\_ack}$**then**
5.   update values of m_{low}, n_{high}, and r_{ave}
6.   if$n_{high} \neq 0$ and $m_{low}/n_{high} < h_d$**then**
7.   send an ACK with (m_{low}, n_{high}, r_{ave})
8.   set $n = 0$.
9.   **else**
10.   set $n = n + 1$
11.   if$n \geqslant d$**then**
12.   send an ACK with (m_{low}, n_{high}, r_{ave})
13.   set $n = 0$.
14.   **end if**
15.   **end if**
16.   set $t_{next\_ack} = t + \tau_i$
17. **end if**
18. **end while**

Packet Allocation

AeroMTP needs a packet allocation scheme to allocate a certain number of encoded packets to each path for a block. Both the total number of encoded packets generated for a block and the distribution of each path should be considered to improve the performance. In Refs.13 and 15, the packet allocation algorithms are designed based on window-based congestion control mechanisms which cannot be directly applied to AeroMTP. Then, a rate-based allocation algorithm needs to be designed for AeroMTP.

In order to increase the goodput, packet allocation algorithm is expected to achieve two objects: low redundancy and high decoding probability. It is very hard to achieve the two objectives at the same time. To achieve a low redundancy, the sender should not transmit redundant encoded packets to receiver once the receiver can recover the original block. However, the sender should send more redundant data through a lossy channel to ensure that the receiver has a high probability of successful decoding. If the receiver is not able to recover the block, it needs to inform the sender to send more encoded packets for the block, which will increase the delivery delay for the block. To handle these problems, the sender is allowed to send some redundant data through lossy channels as long as it can help significantly reduce the delivery delay of a block.

We assume that the sender would like m encoded packets to be successfully delivered to the receiver with a success probability of P_s via a lossy channel where each packet is dropped independently of others with probability p. Let X be the number of packets dropped by the channel before the mth success delivery occurs. Then, we have

$$P\{X = y\} = \binom{m + y - 1}{y}(1 - p)^m p^y \quad y = 0, 1, 2 \ldots$$

$$(3)$$

Using Eq. (3), the mean and variance of random variable X can be easily obtained:

$$E(X) = \frac{mp}{1 - p}$$

$$(4)$$

$$D(X) = \frac{mp}{(1 - p)^2}$$

$$(5)$$

Therefore, the random variable X follows the negative binomial distribution with mean $mp/(1 - p)$ and variance $mp/(1 - p)^2$, i.e., $X \sim NB(m, 1 - p)$. Using the central limit theorem, if m is sufficiently large, X approximately follows normal distribution with mean $mp/(1 - p)$ and variance $mp/(1 - p)^2$, i.e.,

$$X \sim N\left(mp/(1 - p), \quad mp/(1 - p)^2\right) \tag{6}$$

In order to successfully deliver m packets to the receiver with P_s success probability, the average number of packets dropped by the channel, n, satisfies the following requirement:

$$P\{X \leqslant n\} = P_s \tag{7}$$

Using Eqs. (6) and (7) becomes

$$P\left\{\frac{X - \dfrac{mp}{1 - p}}{\sqrt{\dfrac{mp}{1 - p}}} \leqslant \frac{n - \dfrac{mp}{1 - p}}{\sqrt{\dfrac{mp}{1 - p}}}\right\} = P_s \tag{8}$$

In order to satisfy Eq. (8), the value of n should satisfy

$$\frac{n - \dfrac{mp}{1 - p}}{\sqrt{\dfrac{mp}{1 - p}}} = \Phi^{-1}(P_s) \tag{9}$$

Then, n can be calculated by

$$n = \left\lceil \frac{\Phi^{-1}(P_s)\sqrt{mp} + mp}{1 - p} \right\rceil \tag{10}$$

where $\lceil x \rceil$ is the smallest integer larger than x. Therefore, to ensure that m encoded packets can be successfully delivered with P_s success probability, the average number of encoded packets required to be sent via a channel with loss rate p, denoted by $\Psi(m, p)$, can be calculated by

$$\Psi(m, p) = n + m = \left\lceil \frac{\Phi^{-1}(P_s)\sqrt{mp} + mp}{1 - p} \right\rceil + m = \left\lceil \frac{\Phi^{-1}(P_s)\sqrt{mp} + m}{1 - p} \right\rceil \tag{11}$$

If there exists only one path with lossy rate p between the sender and receiver, the sender just needs to send $\Psi(k(1+\varepsilon), p)$ encoded packets for the block containing k symbols, which can achieve satisfactory performance by selecting an appropriate value for the expected success probability P_s. However, in the network with multiple heterogeneous paths, we also need to allocate encoded packets to individual paths such that the receiver can successfully receive $k(1+\varepsilon)$ encoded packets for each block as fast as possible.

For path i, the sender maintains its average transmission rate b_i, the RTT value RTT_i and the loss rate p_i. Suppose that m_i encoded packets of a block are allocated to path i, i.e., m_i encoded packets should be successfully delivered via path i. In order to ensure that the receiver can successfully receive these m_i encoded packets through lossy channels, the number of encoded packets required to be sent via path i is equal to $\Psi(m_i, p_i)$. Let BUF_i be the number of encoded packets for other blocks that have been in the transmission buffer of path i. The delivery time for m_i packets via path i, denoted by $DT_i(m_i)$, is defined as the duration from the moment that $\Psi(m_i, p_i)$ packets are sent to the transmission buffer to the time when the receiver successfully receives m_i packets.

As shown in Fig. 5, $DT_i(m_i)$ consists of three parts: transmission time for BUF_i packets, transmission time for $\Psi(m_i, p_i)$ packets and end-to-end delay of path i, which can be estimated by BUF_i/b_i, $\Psi(m_i, p_i)/b_i$, and $RTT_i/2$ respectively. Therefore, if $b_i \neq 0$, $DT_i(m_i)$ can be estimated as

$$DT_i(m_i) = \frac{\Psi(m_i, p_i) + BUF_i}{b_i} + \frac{RTT_i}{2} \tag{12}$$

If $b_i = 0$, $DT_i(m_i)$ can be calculated as

$$DT_i(m_i) = \begin{cases} 0 & \text{if} \quad m_i = 0 \\ +\infty & \text{if} \quad m_i \neq 0 \end{cases} \tag{13}$$

Assume that there exist N paths between the sender and receiver. The block delivery time is defined as the duration from the moment that the encoded packets of a block are sent to transmission buffers of the N paths to the time when the block is decoded successfully. Since the receiver can recover a block with a very high probability when $k(1+\varepsilon)$ encoded packets of the block are received, the block delivery time can be approximately estimated by the time elapsed before the receiver successfully receives $k(1+\varepsilon)$ encoded packets for the block. Thus, the block delivery time can be determined by the maximum value of $DT_i(m_i)$, $i = 1, 2, \ldots, N$.

In order to increase the goodput, the objective of the packet allocation is to figure out the allocation variables $\{m_1, m_2, \ldots, m_N\}$ such that the value of $\max_{i=1,2,\ldots,N}\{DT_i(m_i)\}$ is minimal. Then, the packet allocation can be modeled as the following integer programming problem:

$$\min_{\{m_1, m_2, \ldots, m_N\}} \max_{i=1,2,\ldots,N}\{DT_i(m_i)\}$$

$$\text{s.t. } \sum_{i=1}^{N} m_i = \lceil k(1+\varepsilon)\rceil$$

$$0 \leqslant m_i \leqslant \lceil k(1+\varepsilon)\rceil, \quad i = 1, 2, , N$$

$$m_i \text{ integer}, \quad i = 1, 2, , N \tag{14}$$

Since N is not very large (typically 2–4 in ANs), the approximate solution of the optimization problem can be quickly obtained by some heuristic algorithms. Based on the allocation variables $\{m_1, m_2, \ldots, m_N\}$ calculated by Eq. (14), the sender needs to generate $\sum_{i=1}^{N} \Psi(m_i, p_i)$ encoded packets for the block and transmits $\Psi(m_i, p_i)$ encoded packets via path i.

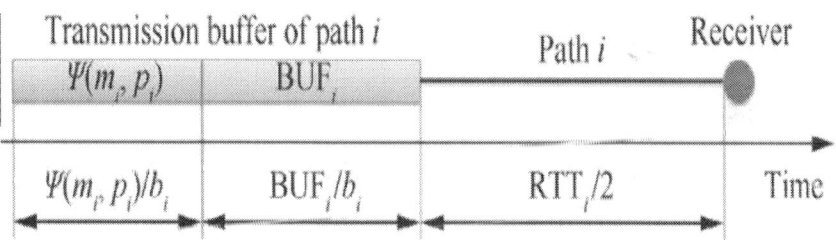

Figure 5: Delivery time for m_i packets via path i.

Once there exists a path with empty transmission buffer and not in the blackout state, the sender allocates and transmits encoded packets for the next available block. To determine if the next block is available to be sent, a block-level sliding window mechanism is applied at the sender. The window size gives the maximum number of unacknowledged blocks, which can be set to an arbitrary value. Once a block is successfully recovered by the receiver, it returns an acknowledgment to the sender, otherwise, sends feedback to the sender to acquire additional encoded packets for the block. If a block is acknowledged, the sliding window shifts by one unit and the next block is available to be sent.

SIMULATION RESULTS

In this section, the performance of AeroMTP is evaluated using OMNeT++.[27] Firstly, we set appropriate values for protocol parameters. Secondly, the packet redundancy is analyzed and compared. Thirdly, we investigate the performance of the rate-based congestion control scheme in AeroMTP. Finally, the performance of AeroMTP is analyzed in heterogeneous multihoming networks.

Parameter Setting

Parameters of AeroMTP should be determined appropriately to ensure a good performance. The block size k is an important parameter, which can influence the performance of fountain code. For a fixed value of ε, a large block size k can make the decoding failure probability low when $k(1 + \varepsilon)$ encoded packets of a block are received. [16]However, a large block size also brings about a long block delivery delay and a small block size is preferred for some real-time applications. Thus, we should choose an appropriate value for k according to the real-time requirements of applications. For simplicity, we set $k = 1000$ in the simulation experiments. As described in Section 4.2, the receiver counts the number of received packets from path i in a sliding time window of T_i and sends back ACKs every τ_i period. In order to accurately update the transmission rate every RTT_i period at the sender side, T_i should be larger than RTT_i. Thus, the value of τ_i and T_i should be set according to RTT_i and we set $\tau_i = 0.5RTT_i$ and $T_i = 2RTT_i$ in the simulations. The protocol parameters given in Table 3 are used during simulation experiments unless otherwise stated.

Table 3: Protocol parameters used in experiments.

Parameter	Value	Definition
P_s	0.92	Success probability
k	1000	Block size
ε	0.05	Coding redundancy
τ_i	$0.5RTT_i$	ACK period for path i
d	1	Delaying factor
T_i	$2RTT_i$	Sliding time window of path i for packets counting
h_d	0.8	Rate decrease threshold

Redundancy Analysis

We define packet redundancy as the fraction of additional packets generated to ensure that a certain number of packets can be successfully delivered via a lossy channel. If the sender would like m packets to be successfully delivered via a channel with loss rate $p,m + n$ packets are generated and sent to the receiver, and then the packet redundancy is n/m. It is crucial to determine the packet redundancy for each block according to the loss rate. If the packet redundancy is too small, the receiver cannot recover the original block with a high probability. If the redundancy is too large, the bandwidth is wasted. In MPLOT, [13] the packet redundancy is calculated by $\delta p/(1 - p)$, where parameter δ is determined by the expected success probability P_s and the values of δ for different P_s are listed in Ref. [13]. Obviously, according to Eq. (11), the packet redundancy in AeroMTP can be written as $(\Psi(m, p) - m)/m$. Here, we set $m = 1000$ and compare the packet redundancy between MPLOT and AeroMTP. Fig. 6(a) plots the packet redundancy as a function of loss rate for MPLOT and AeroMTP when $P_s = 0.99$. We also investigate the redundancy difference between MPLOT and AeroMTP which is obtained by subtracting the redundancy of AeroMTP from that of MPLOT. Fig. 6(b) shows the redundancy difference as a function of loss rate when P_s is set to 0.8 and 0.99 respectively. As shown in Fig. 6, AeroMTP can achieve observably lower redundancy than MPLOT as the loss rate increases, which means that smaller number of packets is needed to be sent by AeroMTP than MPLOT to ensure a certain number of packets can be successfully delivered via a channel with high loss rate. Then, we consider a simple scenario where a sender sends data to a receiver via a lossy channel. In the scenario, the data is divided into blocks with the size of 1000, and the sender converts each block to a certain number of encoded packets which is then sent to the receiver. Here, the coding redundancy is not considered, i.e., a block can be recovered by the receiver if at least 1000 encoded packets for the block is successfully received. We implement the experiments to investigate the success probability of block recovery when the number of encoded packets generated for each block is determined by AeroMTP and MPLOT respectively.Fig. 7 illustrates the success probability as a function of loss rate when the expected success probability P_s is set to 0.8 (see Fig. 7(a)) and 0.99 (see Fig. 7(b)). We observe inFig. 7 that there is a certain gap between the actual success probability achieved by MPLOT and its expected probability while AeroMTP roughly achieves the expected success probability at different loss rates. At high loss rate, i.e., $p > 0.1$, MPLOT generates excessive packet redundancy and achieves much higher success probability than the expected value. Moreover, when packet loss rate is less than about 0.07, the redundancy of MPLOT is inadequate which leads to a low success probability. Unlike MPLOT, AeroMTP well achieves the expected success probability, irrespective of the packet loss rate. As shown in Figs. 6 and 7, AeroMTP can well realize the expected success probability with lower packet redundancy than MPLOT. The main reason is that we obtain the closed-form expression for the packet redundancy and it is more accurate than that in MPLOT which is obtained just through curve-fitting.

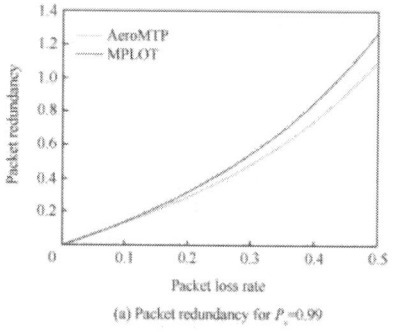
(a) Packet redundancy for $P_s = 0.99$

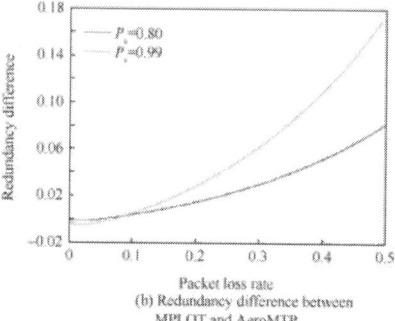
(b) Redundancy difference between MPLOT and AeroMTP

Figure 6: Redundancy comparison.

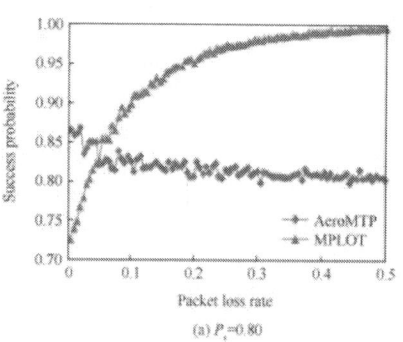
(a) $P_s = 0.80$

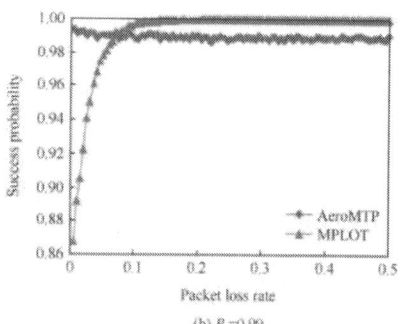
(b) $P_s = 0.99$

Figure 7: Success probability of block recovery.

Success probability P_s is an important parameter in AeroMTP which should be set to an appropriate value. Intuitively, the protocol can get a good performance when a high P_s is used. However, a too high P_s needs too large amount of additional packets which will result in bandwidth wastage. Fig. 8 shows the packet redundancy in AeroMTP as a function of P_s when $m = 1000$, $p = 0.05$ and 0.20 respectively. We can see that the packet redundancy almost linearly increases with P_s when $P_s < 0.92$, and then rises sharply. This means that much more additional packets are needed to ensure that m packets are successfully delivered with probability $P_s > 0.92$. Thus, the protocol with $P_s > 0.92$ are considered not worth the additional costs, and then we set $P_s = 0.92$ in the simulation experiments

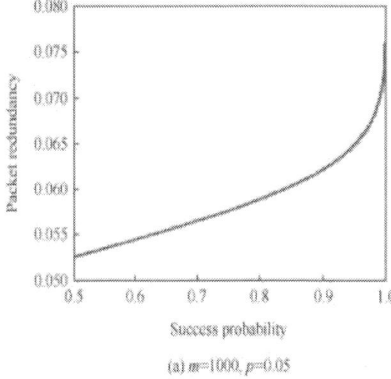
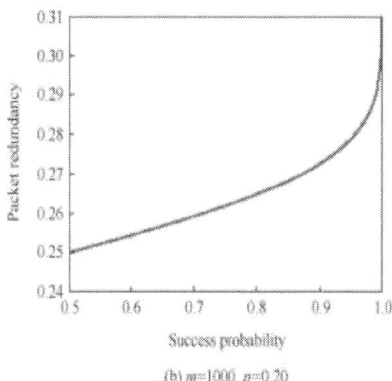

(a) $m=1000, p=0.05$ (b) $m=1000, p=0.20$

Figure 8: Packet redundancy as a function of success probability.

Performance of Rate-Based Congestion Control Scheme

In this section, we investigate the performance of the rate-based congestion control scheme in AeroMTP. A single-path network model is established to study the performance. We first investigate throughput performance of the congestion control scheme under different channel conditions and then analyze the fairness performance for competing flows. Finally, we consider a asymmetric path to evaluate the improvement achieved by the delaying factor.

Network Model

In order to investigate the performance of the rate-based congestion control scheme in AeroMTP, we first simulate a network topology where N sources, $S_1, S_2, \dots, S_N$, transmit data to N destinations, $D_1, D_2, \dots, D_N$, using AeroMTP (see Fig. 9). Since there exists only one path in a source–destination pair, AeroMTP used in this scenario can be called single-path AeroMTP. The N flows are multiplexed in the router A and the multiplexed streams are transmitted to the router B via a lossy channel with a capacity of 500 packets/s and a loss rate of p. The multiplexed streams are de-multiplexed in the router B and then routed to the corresponding N destinations. For simplicity, we assume that there is no blackout along the path during the experiments. The routers are equipped with a simple priority-queuing mechanism with two priority levels (low and high priority), following first-in-first-out (FIFO) queuing with priority-based preemption. In this priority-queuing mechanism, low priority packets are discarded first in the case of congestion when the queue is overutilized and overflown. Each router has a buffer that can accommodate 50 packets.

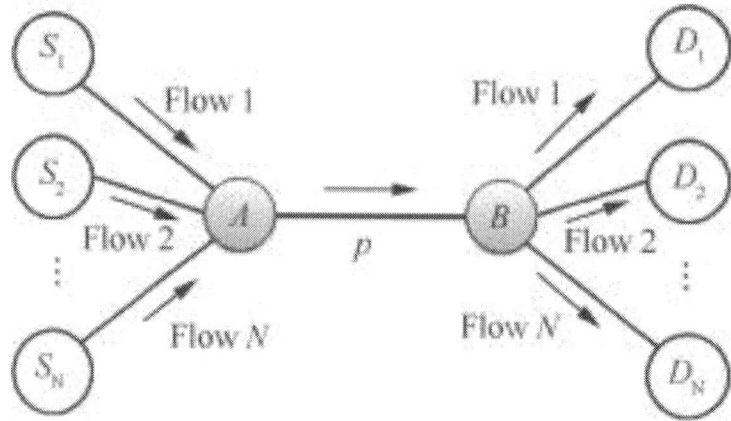

Figure 9: Single-path network topology.

Throughput Performance

We first set $N = 1$ and investigate the average throughput achieved by the flow for AeroMTP and traditional TCP with SACK option respectively. Fig. 10(a) shows the average throughput as a function of packet loss rate when RTT = 80 ms, and Fig. 10(b) plots the average throughput for different RTT values when the channel has a loss rate of 0.01. We can see that the congestion control scheme of AeroMTP can achieve a better performance than TCP, especially when the path has high loss rates and RTT values. The main reason is that the congestion control mechanism in TCP detects congestion depending on packet loss, which are unable to distinguish between losses due to channel errors and congestion in the lossy channel. When packet loss occurs due to link errors, the TCP source also decreases its transmission rate. Then, the network efficiency decreases drastically for a network with high loss rate and this problem is amplified by high RTT values. AeroMTP can decouple congestion decisions from packets losses and avoid the erroneous congestion decisions due to high link errors.

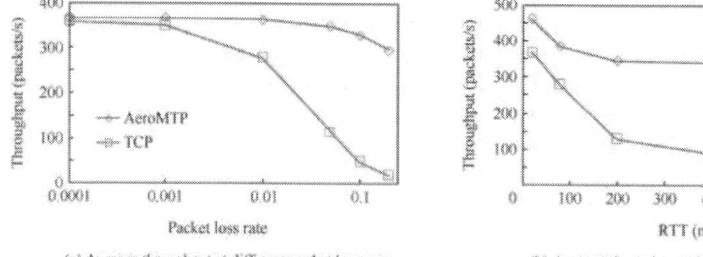

Figure 10: Average throughput of flow.

Then, we set $N = 2$ and investigate the throughput achieved by competing flows for AeroMTP. We consider a network at high loss rate, i.e., $p = 0.2$, and plot instantaneous throughput of the two AeroMTP flows in Fig. 11 when the RTT has a low (50 ms), high (500 ms) and volatile (RTT changes uniformly between 70 ms and 300 ms every 2 s) value respectively. As shown in Fig. 11, we can see that all the AeroMTP flows can achieve a stable throughput for different RTT values and share the bandwidth equally.

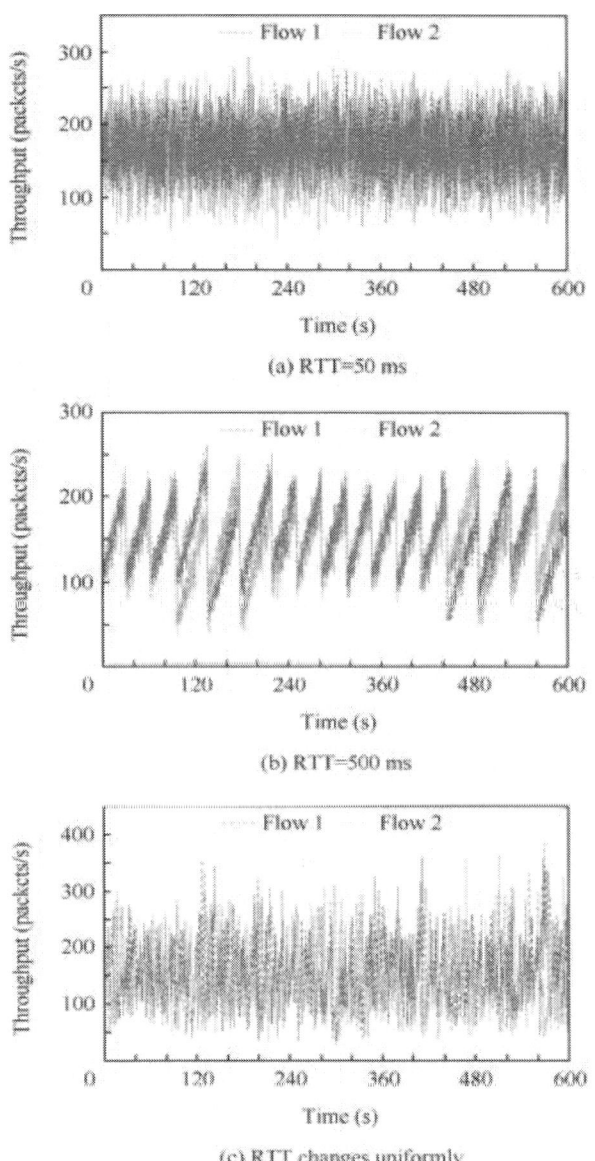

Figure 11: Instantaneous throughput of flows.

Fairness Analysis

In ANs, multiple independent communication paths can be simultaneously established between any two nodes via the communication relays and MANET. Due to the special nature of MANET, it is difficult to analyze fairness for the paths via MANET. In most cases, ANs communicate mainly via relays. Thus, it is necessary to investigate the fairness performance for AeroMTP. For the fairness performance of the congestion control mechanism in AeroMTP, we consider two different scenarios, i.e., homogeneous and heterogeneous fairness. To assess the fairness, we use Jain's fairness index[28] which quantifies the degree of similarity between the amount of link resources used by all connections. Let x_i be the throughput of the ith connection and let s be the number of connections competing for the same bottleneck resources. Then, the Jain's fairness index, J, can be evaluated as

$$J = \frac{\left(\sum_{i=1}^{s} x_i\right)^2}{s \sum_{i=1}^{s} x_i^2}$$

(15)

when the fairness index is 1, all of the connections consume the same amount of bottleneck resources. The fairness index J decreases as the difference between the throughput values achieved by different connections increases.

We first consider the homogeneous fairness, i.e., the fairness of the 10 single-path AeroMTP connections. Fig. 12(a) shows the fairness index as a function of packet loss rate when the RTT has a low (50 ms) and high (500 ms) value. We can see that the AeroMTP can have a good fairness performance for different RTT and packet loss rate in the homogeneous scenario.

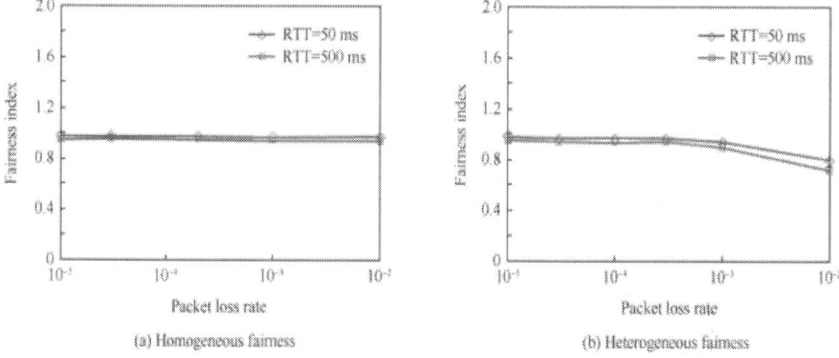

(a) Homogeneous fairness   (b) Heterogeneous fairness

Figure 12: Fairness index as a function of packet loss rate.

Next, we investigate the heterogeneous fairness, i.e., the fairness of the five single-path AeroMTP connections and five TCP connections. As shown in Fig. 10, the performance of TCP degrades significantly due to high link errors, which can significantly impact the fairness investigation. Thus, we cannot use the experiment scenario that both the TCP and AeroMTP connections pass through the same type of lossy link. Similar to Ref.[29], we assume that TCP packets are not dropped when they pass through the lossy channel.Fig. 12(b) shows the fairness index between AeroMTP and TCP connections, where TCP packets are not lost in the channel. Since the throughput performance of AeroMTP degrades with increasing packet loss rate and the TCP connections are not influenced by packet loss, the fairness index decreases when the packet loss rate is getting high. Nevertheless, the fairness index is still approximately 1 for different RTTs and packet loss rates. Furthermore, the fairness index shown in Fig. 12(b) is in close proximity to that in Ref.[29] which uses the same approach to investigate the heterogeneous fairness. Therefore, AeroMTP follows TCP-friendly rules and can fairly share the link resource with other TCP-friendly transport protocols.

Performance Analysis in Bandwidth Asymmetry Path

AeroMTP uses the delaying factor to address bandwidth asymmetry problems in ANs. We set $N = 1$ and set the reverse link capacity to 5 packets/s, i.e., the bandwidth asymmetry is 100:1, to show the performance improvement achieved by delaying factor. In the experiments, RTT changes uniformly between 50 ms and 150 ms every second, and the loss rate changes uniformly between 0.001 and 0.1 every second. Here, we assume that both encoded packets and ACKs have the same size of 1 KB, and use the goodput (i.e., the number of useful bytes transferred per second) as the performance metric. Fig. 13 plots the goodput achieved by the flow for different values of delaying factor d. We can see that an increase in the delaying factor also leads to an increase in the goodput. However, goodput decreases for both cases with either too high or too small delaying factor of d. This is because if it is too small, reverse link is congested, resulting in massive ACKs lost in the reverse path. Then, the source may not detect the congestion in the forward link, which results in massive encoded packets lost due to congestion in the forward link and then causes goodput decrease. On the other hand, if d is too large, the source receives less number of ACKs than it expects, which influences the estimation of RTT and packet loss rate for the path and leads to performance degradation. As shown inFig. 13, goodput achieves the peak value at $d = 4$. In the experiments, the optimal value of d is found manually, but it is a possible option to perform this task by an adaptive algorithm of the protocol, so it will be part of our future research.

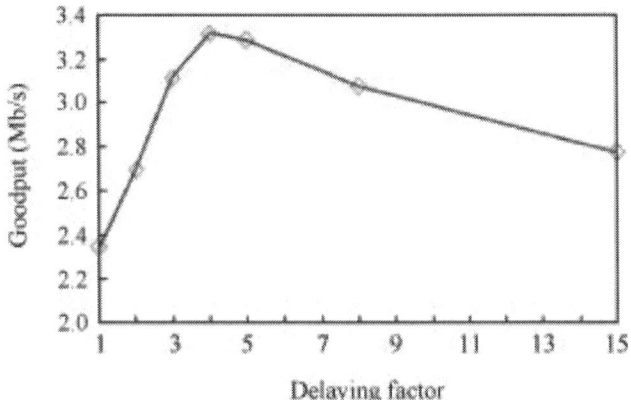

Figure 13: Goodput as a function of delaying factor.

Performance of AeroMTP in Heterogeneous Multihoming Networks

In this section, we investigate the performance of AeroMTP in heterogeneous multihoming networks. Firstly, a network model is established to simulate the heterogeneous multihoming networks. Secondly, we evaluate the performance of AeroMTP when different packet allocation schemes are used. Finally, we implement a network model to simulate the real ANs and investigate the performance of AeroMTP in ANs.

Network Model

In order to investigate the performance of AeroMTP in heterogeneous multihoming networks, we implement a network model to simulate the heterogeneous multihoming networks. We employ a topology with two nodes connected by N disjoint paths (see Fig. 14), in which the term "disjoint" means that the path parameters (i.e., bandwidth, delay, and loss rate) of the N paths vary independently of each other.

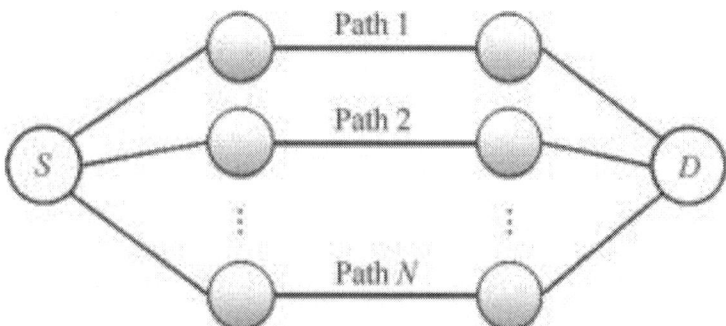

Figure 14: Heterogeneous network topology.

Each path is implemented as a three-state time-varying process (see Fig. 15). Path parameters vary between three different states, i.e., good state, normal state, and bad state, and the time duration between state transitions is exponentially distributed with the mean of D_s. The probability of a transition from a state to any other states is kept at 0.5. Then, we can realize a special dynamic path by adjusting D_s and parameters in each state.

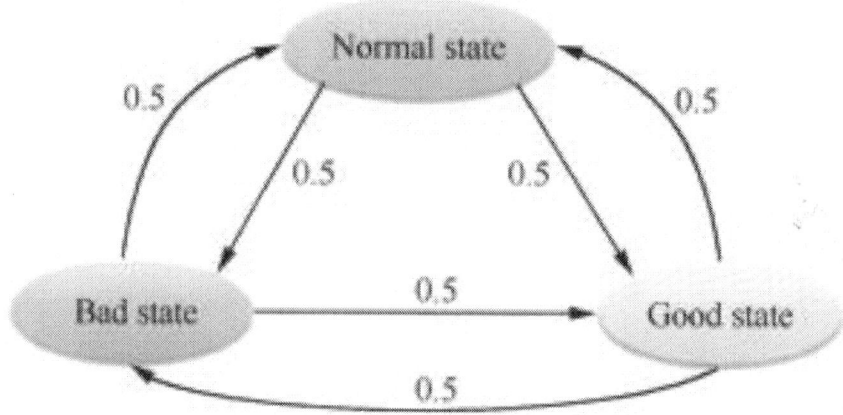

Figure 15: Three-state process model for paths.

Performance Comparison between Different Packet Allocation Schemes
For a multipath transport protocol, the packet allocation scheme can significantly impact the performance. In order to investigate the performance impact of the packet allocation scheme, two allocation schemes are used in the protocol:

(1) Simple scheme: As described in Ref.[13], this scheme estimates the aggregate loss rate across all paths, and determines the total number of encoded packets transmitted for a block. Then, these packets are simply sent to better paths, i.e., paths that have high bandwidth, low loss rate, and short RTT.
(2) Rate-based scheme: As described in Section 4.3, the rate-based scheme estimates packet loss rate for each path, and allocates encoded packets of a block to each path based on the integer programming Eq. (14).

In the experiments, the total capacity across the N paths is set to a constant (8 Mbps) and the bandwidth of a path in the three states is fixed at $8/N$ Mbps. The size of an encoded packet is set to 1 KB and the value of D_s is set to 500 ms.

We first investigate the goodput and average block delivery delay when there exist different numbers of paths between the sender and receiver. The average block delivery delay is the average duration from the transmission of the first encoded packets for a block to the time when this block is decoded successfully. We consider a scenario where loss rate on a path varies between 0.01 (good state), 0.08 (normal state) and 0.15 (bad state), and the delay of a path varies between 20 ms (good state), 40 ms (normal state), and 90 ms (bad state). We observe in Fig. 16 that AeroMTP using the rate-based scheme can achieve shorter average block delivery delay and higher goodput than that using the simple scheme, especially when the number of paths is more than three. According to the integer programming Eq. (14), the rate-base scheme can find approximately optimum solution to minimize block delivery delay, while the simple scheme intuitively selects the best path to send the packets. Consequently, the rate-based scheme can work better than the simple scheme in heterogeneous multihoming networks.

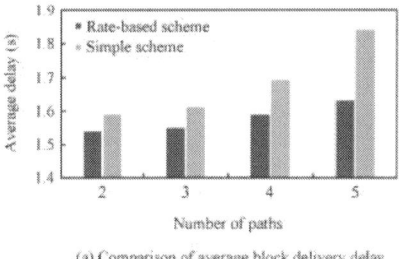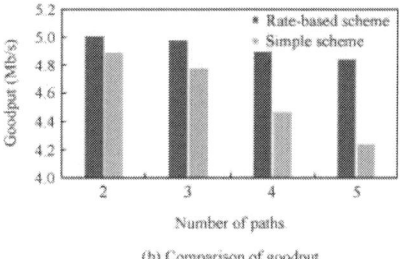

(a) Comparison of average block delivery delay   (b) Comparison of goodput

Figure 16: Performance comparison when different schemes are used for packet allocation.

Next, we assume that there exist four paths between the sender and receiver and investigate the goodput performance for the two packet allocation schemes under different path conditions. We assume that the loss rate and RTT of the four paths vary with the mean of P_{ave} and T_{ave} respectively, i.e., the loss rate varies between $0.5P_{ave}$(good state), P_{ave} (normal state) and $1.5P_{ave}$ (bad state), and the RTT varies between $0.5T_{ave}$ (good state), T_{ave} (normal state) and $1.5T_{ave}$ (bad state). We study the ratio of goodput achieved by rate-based scheme and simple scheme under different path conditions by adjusting the value of P_{ave} and T_{ave}. Fig. 17 plots the goodput ratio as a function of T_{ave} for $P_{ave} = 0.001$ and $P_{ave} = 0.100$ respectively. As shown in Fig. 17, thanks to the optimization in the rate-based scheme, the rate-based scheme achieves higher goodput than the simple scheme under different path conditions, and significantly outperforms the simple scheme for paths with high loss rate and RTT value.

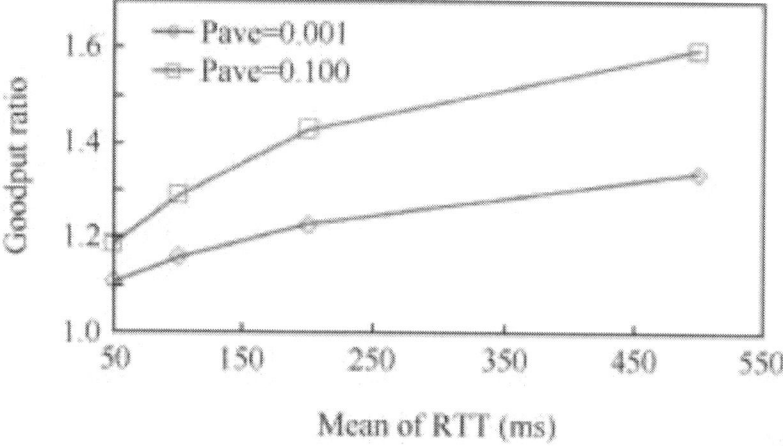

Figure 17: Goodput ratio as a function of T_{ave}.

Performance Analysis in ANs

In order to investigate the performance of AeroMTP in ANs, we should implement a network model to simulate the real ANs. As described in Section 3, three types of communication paths can be simultaneously established between any two nodes of ANs, i.e., paths via communication relays, paths via the MANET formed by omnidirectional links, and paths via the MANET formed by directional links. These three types of paths represent three typical channels in ANs. Since the communication relay has limited bandwidth resource and long propagation time, communication paths via relays are characterized by low bandwidth and long delay. Paths via the MANET formed by omnidirectional links have higher bandwidth and lower delay than those via a relay. However, the path conditions (bandwidth, delay, and loss rate) change frequently due to the highly dynamic MANET. Paths via the MANET formed by high bandwidth directional links have the highest bandwidth and lowest delay than the other two types of paths. Compared with omnidirectional links, directional links are more likely to be influenced by aircraft movements and weather conditions, which makes path conditions have the highest dynamics. Then, we implement a network model with three paths to simulate the three types of communication paths in ANs. The parameters for three paths are set in Table 4. In Table 4, Path 1, 2, and 3 represent the paths via communication relays, paths via MANET formed by omnidirectional links, and paths via MANET formed by directional links respectively. We use two variables P_{loss} and D_m to indicate the loss rate and dynamics of the three paths, which allows us to investigate the protocol's performance under different conditions. For simplicity, we assume that the three paths have the same loss rate values, i.e., $0.5P_{loss}$, P_{loss}, and 1 for good state, normal state, and bad state respectively. If a path transits to a state where loss rate is equal to 1, we can also say that the path transits to the blackout state. The packet size for all packets is also set to 1 KB.

Table 4: Parameter setting.

Path	Parameter	Good state	Normal state	Bad state	D_s
	Bandwidth	60 packets/s	50 packets/s	30 packets/s	
1	Delay	0.19 s	0.2 s	0.21 s	D_m
	Loss rate	$0.5P_{loss}$	P_{loss}	1 (BLACKOUT)	
	Bandwidth	200 packets/s	150 packets/s	130 packets/s	
2	Delay	0.06 s	0.08 s	0.12 s	$0.5D_m$
	Loss rate	$0.5P_{loss}$	P_{loss}	1 (BLACKOUT)	
	Bandwidth	800 packets/s	600 packets/s	400 packets/s	
3	Delay	0.01 s	0.04 s	0.06 s	$0.1D_m$
	Loss rate	$0.5P_{loss}$	P_{loss}	1 (BLACKOUT)	

Take Path 2 as example, the instantaneous transmission rate and ratio of received low- and high-priority packets are plotted in Fig. 18. We can see that the sender can well adjust the transmission rate according to the path's actual bandwidth by controlling the ratio of received low- and high-priority packets at about 1. For traditional window-based congestion control algorithms, path's frequent blackouts make the protocol go to slow start phase frequently, leading to a severe reduction of throughput. As shown inFig. 18(a), the transmission rate is set to 0 when blackout is detected and rapidly recovers as soon as the blackout is over, which can significantly reduce the impact of frequent blackouts on the throughput.

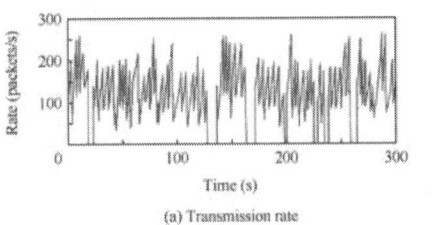

(a) Transmission rate

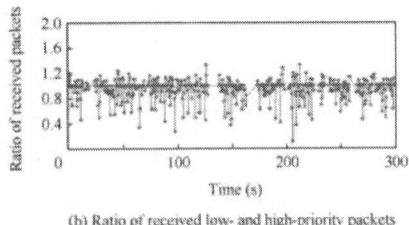

(b) Ratio of received low- and high-priority packets

Figure 18: Instantaneous value of H and transmission rate in Path 2.

In order to investigate the performance improvement by fountain code, we also implement AeroMTP without the encoding procedure, called non-coding AeroMTP, i.e., the k symbols of a block with a size of k are directly packetized and allocated to different paths. Then, we set $D_m = 1$ s and compare the goodput performance of AeroMTP with that of non-coding AeroMTP in the scenario as shown in Table 4. Fig. 19 shows the goodput achieved by AeroMTP and non-coding AeroMTP for the network withP_{loss} = 0.001, 0.01 and 0.1 respectively. We can see that the goodput performance is improved significantly for the network with high loss rate when fountain code is employed as packet-level FEC code. Non-coding AeroMTP needs to retransmit the packets which are lost in the channel, resulting in frequent retransmission and performance degradation in the network at high loss rate. AeroMTP employs fountain code as packet-level FEC code to recover from data loss and avoid retransmissions, which can significantly improve the performance for the network at high loss rate.

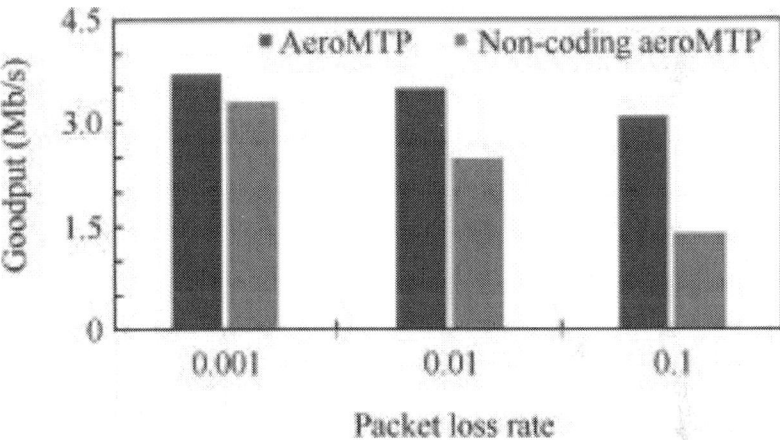

Figure 19: Goodput comparison between AeroMTP and non-coding AeroMTP for different values of P_{loss}.

Then, we compare the performance of AeroMTP with another fountain code-based multipath transport protocol FMTCP[15] in the scenario as shown in Table 4. In order to compare the two protocols under different conditions, we investigate the performance when paths have high (D_m = 1 s) and low (D_m = 10 s) dynamics. We first investigate the average packet redundancy for AeroMTP and FMTCP when the loss variable P_{loss} has different values. The ratio between average packet redundancy for AeroMTP and FMTCP is plotted in Fig. 20. We can observe in Fig. 20 that the average packet redundancy in AeroMTP is larger than that in FMTCP, which means that AeroMTP brings more overhead than FMTCP, especially for the channels with high loss rates and high dynamics. This is because FMTCP uses window-based congestion control and sends the encoded packets of a block in several bursts. In FMTCP, the sender decides how many more encoded packets should be sent for the block according to the ACK of previously sent packets. Thus, packets lost except those sent during the last RTT can be detected and supplemented in time, and FMTCP only needs to append extra packets for the encoded packets of the block sent during the last RTT, which will be no more than the congestion window size. In AeroMTP, all the encoded packets for a block are generated before they are sent to the channels and this behavior makes a more pessimistic estimation than FMTCP. Therefore, AeroMTP sends more encoded packets for a block than FMTCP, which results in a larger packet redundancy. Next, we also investigate the goodput and standard deviation of block delivery delay for AeroMTP and FMTCP.Fig. 21(a) illustrates the ratio between standard deviation of block delivery delay in AeroMTP and FMTCP. We observe in Fig. 21(a) that block delivery delay in AeroMTP is more stable than that in FMTCP, which indicates that AeroMTP achieves more stable goodput than FMTCP. Fig. 21(b) plots the ratio between goodput achieved by AeroMTP and FMTCP for the network with different P_{loss} and different dynamics. As shown in Fig. 21, although the average packet redundancy in AeroMTP is larger than that in FMTCP, AeroMTP achieves a higher and more stable goodput than FMTCP, especially when the communication paths have high loss rates and high dynamics. There are two main reasons. On the one hand, AeroMTP employs a rate-based congestion control mechanism which can work well in the channels with high loss rate and frequent blackouts, while FMTCP uses the conventional window-based congestion control which will significantly degrade the throughput of the channels. On the other hand, FMTCP allocates encoded packets to paths only based on the expected packet arrival time over different paths, and thus it is difficult for FMTCP to find the optimal allocations. According to the integer programming Eq. (14), AeroMTP can always find the approximately optimum solution for packet allocation and achieves a low and stable block delivery delay. Consequently, at the cost of large packet redundancy, AeroMTP increases the throughput of each path and minimizes block delivery delay, which results in a higher and more stable goodput than FMTCP.

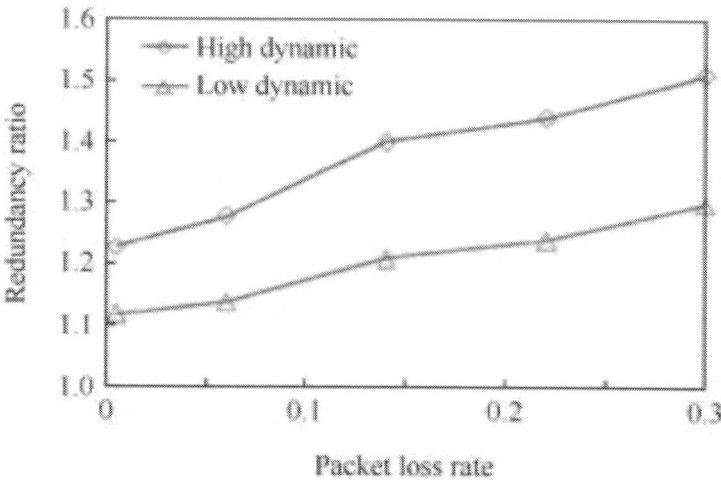

Figure 20: Redundancy ratio as a function of P_{loss}.

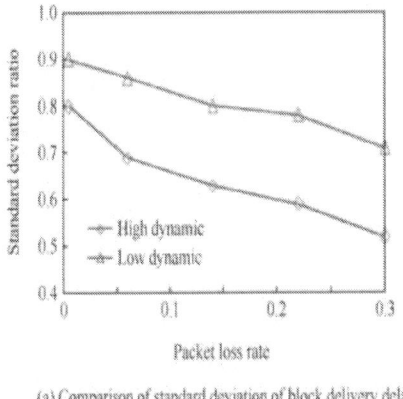
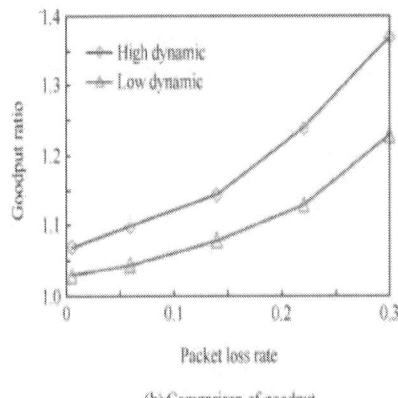

(a) Comparison of standard deviation of block delivery delay

(b) Comparison of goodput

Figure 21: Performance ratio as a function of P_{loss}.

CONCLUSIONS

(1) In this paper, we propose AeroMTP, a fountain code-based multipath transport protocol that can effectively utilizes multiple heterogeneous paths in ANs to improve the network performance.

(2) Fountain code is employed as packet-level FEC code to recover from data loss and avoid retransmissions in AeroMTP.

(3) A TCP-friendly rate-based congestion control mechanism is deployed for each path, which decouples congestion decisions from packets losses in order to avoid the erroneous congestion decisions due to high link errors. By adjusting the delaying factor, the amount of traffic on the reverse channel can be controlled, which well solves the bandwidth asymmetry problem in ANs.

(4) In order to reduce the effects of blackout conditions on the throughput performance of each path, AeroMTP incorporates the blackout state procedure into the protocol operation.

(5) A packet allocation algorithm is designed based on optimization to minimize the block delivery time and consequently improve the total goodput over all paths.

(6) Simulations show that AeroMTP outperforms other multipath transport protocols in ANs and is of great potential to be applied to ANs to enhance the network performance.

REFERENCES

1. Wang Y. Fundamental issues in systematic design of airborne networks for aviation. Proceedings of IEEE aerospace conference; Big Sky, MT, USA. Piscataway, NJ: IEEE Press; 2006. p. 1–8.

2. Karras K, Kyritsis T, Amirfeiz M. Aeronautical mobile ad hoc networks. Proceedings of 14th European wireless conference; Hong Kong, China. Piscataway, NJ: IEEE Press; 2008. p. 3972–7.

3. Schnell M, Scalise S. NEWSKY: a concept for networking the sky for civil aeronautical communications. Space Commun 2008;21(3):157–66.

4. Kwak KJ, Sagduyu Y, Deng J, Yackoski J, Li J. Airborne network evaluation: challenges and high fidelity emulation solution. Proceedings of ACM MobiHoc workshop on airborne networks and communications. New York, USA. New York: ACM Press; 2012. p. 49–52.

5. Airborne network architecture – system communication description and technical architecture profile (version 1.1). White – Paterson, Ohio, USA: USAF Airborne Network Special Interest Group; 2004.

6. Sharma V, Subramanian V, Kar K, Kalyanaraman S. A multipath transport protocol to exploit network diversity in airborne networks. Proceedings of the military communications conference; 2008 Nov.16–19; Washington, D.C, USA. Piscataway, NJ: IEEE Press; 2008. p. 1–7.

7. Barre′ S, Paasch C, Bonaventure O. Multipath TCP: from theory to practice. Proceedings of the 10th international IFIP TC 6 networking conference; Valencia, Spain. Berlin: Springer ; 2011. p. 444–57.

8. Rohrer JP, Jabbar A, Cetinkaya EK, Perrins E, Sterbenz JPG. Highly-dynamic cross-layered aeronautical network architecture. IEEE Trans Aerosp Electron Syst 2011;47(4):2742–65.

9. Rohrer JP, Perrins E, Sterbenz JPG. End-to-end disruptiontolerant transport protocol issues and design for airborne telemetry networks. Proceedings of the international telemetering conference; San Diego, CA, USA. San Diego, CA: ITC; 2008. p. 1–10.

10. Stewart R. Stream control transmission protocol. RFC4960 [Internet]. 2007 [cited 2014 August 15]; Available from: <http://tools.ietf.org/html/rfc4960>.

11. Iyengar J, Amer P, Stewart R. Concurrent multipath transfer using SCTP multihoming over independent end-to-end paths. IEEE/ACM Trans Networking 2006;14(5):951–64.

12. Ford A, Raiciu C, Handley M, Bonaventure O. TCP extensions for multipath operation with multiple addresses. RFC6824 [Internet]. 2013 [cited 2014 August 20]; Available from: <http://tools.ietf.org/html/rfc6824>.

13. Sharma V, Kar K, Ramakrishnan KK, Kalyanaraman S. A transport protocol to exploit multipath diversity in wireless networks. IEEE/ACM Trans Networking 2012;20(4):1024–39.

14. Hwang Y, Saha A, Choi H, Lim H, Obele BO. HMTP: multipath transport protocol for multihoming wireless erasure networks. Trans Emerg Telecommun Technol 2014. http://dx.doi.org/ 10.1002/ett.2809.

15. Cui Y, Wang L, Wang X, Wang H, Wang Y. FMTCP: a fountain code-based multipath transmission control protocol. IEEE/ACM Trans Networking 2015;23(2):465–78.

16. MacKay DJC. Fountain codes. IEE Proc Commun 2005;152(6):1062–8.

17. Wang R, Gutha B, Rapet PV. Window-based and rate-based transmission control mechanisms over space-internet links. IEEE Trans Aerosp Electron Syst 2008;44(1):157–70.

18. Luby M. LT Codes. Proceedings of the 43rd annual IEEE symposium on foundations of computer science; Vancouver, BC, Canada. Piscataway, NJ: IEEE Press; 2002. p. 271–80.

19. Shokrollahi A. Raptor codes. IEEE Trans Inf Theory 2006;52(6):2551–67.

20. Luby M, Shokrollahi A, Watson M, Stockhammer T. Raptor forward error correction scheme for object delivery. RFC5053 [Internet]. 2007 [cited 2014 August 25]; Available from: <http://tools.ietf.org/html/rfc5053>.

21. Luby M, Shokrollahi A, Watson M, Stockhammer T, Minder L. RaptorQ forward error correction scheme for object delivery. RFC6330 [Internet]. 2011 [cited 2014 August 25]; Available from: http://tools.ietf.org/html /rfc6330.

22. Ramakrishnan K, Floyd S, Black D. The addition of explicit congestion notification (ECN) to IP. RFC3168 [Internet]. 2001 [cited 2014 August 25]; Available from: <http://tools.ietf.org/ html/rfc3168>.

23. Ba´guena M, Toh CK, Calafate CT, Cano JC, Manzoni P. RCDP: raptor-based content delivery protocol for unicast communication in wireless networks for ITS. J Commun Networks 2013;15(2):198 206.

24. Fang J, Akyildiz IF. RCP-planet: a rate control protocol for interplanetary internet. Int J Satell Commun Networking 2007;25(2):167–94.

25. Akan O¨B, Fang J, Akyildiz IF. TP-planet: a reliable transport protocol for interplanetary internet. IEEE J Select Areas Commun 2004;22(2):348–61.

26. Rejaie R, Handley M, Estrin D. RAP: an end-to-end rate-based congestion control mechanism for realtime streams in the internet. Proceedings of the 18th annual joint conference of the IEEE computer and communications societies.1995 Mar 21–25; New York, NY, USA. Piscataway, NJ: IEEE Press; 1999. p. 1337–45.

27. OMNeT++ Internet. 2014 [cited 2014 August 27]; Available from: <http://www.omnetpp.org>.

28. Chiu D, Jain R. Analysis of the increase and decrease algorithm for congestion avoidance in computer networks. Comput Networks ISDN Syst 1989;17(1):1–14.

29. Akyildiz IF, Akan O"B, Morabito G. A rate control scheme for adaptive real-time applications in IP networks with lossy links and long round trip times. IEEE/ACM Trans Networking 2005;13(3):554–67.

CITATION

Jie Li, Erling Gong, Zhiqiang Sun, Wei Liu, Hongwei Xie, AeroMTP: A fountain code-based multipath transport protocol for airborne networks, Chinese Journal of Aeronautics, Volume 28, Issue 4, August 2015, Pages 1147-1162, ISSN 1000-9361, http://dx.doi.org/10.1016/j.cja.2015.05.010.

CHAPTER 4

Design for Aircraft Engine Multi-Objective Controllers with Switching Characteristics

Liu Xiaofeng [a], *Shi Jing* [a], *Qi Yiwen* [b] , *Yuan Ye* [c]

[a] School of Transportation Science and Engineering, Beihang University, Beijing 100191, China
[b] School of Automation, Shenyang Aerospace University, Shenyang 110136, China
[c] School of Energy and Power Engineering, Beihang University, Beijing 100191, China

ABSTRACT

The aircraft engine multi-loop control system is described and the switching control theory is introduced to solve the regulating and protecting control problems in this paper. The aircraft engine multi-loop control system is firstly described and the control problems are formulated. Secondly, the theory of the smooth switching control is devoted and a new extended scheme for the smooth switching of a switched control system is introduced. Then, for the key technologies of aero-engines switching control, a design algorithm is presented which can determine which candidate controller should be put in feedback with the plant to achieve a desired performance and the procedure to design the aircraft engine multi-loop control system is detailed. The switching performance objectives and the switching scheme are given and a family of PID controllers and compensators is designed. The simulation shows that using the switching control design method can not only improve the dynamic performance of the aircraft engine control system and reduce the switching times, but also guarantee the stability in some peculiar occasions.

INTRODUCTION

With the development of advanced aircraft engine technology, the control technology plays an increasingly important role. Because the aircraft engines have become more complex, with more control signals and higher demands on performance and functionality, electronic control systems have been introduced.1 and 2 The experience of development of the aircraft engines in the world shows that the performance parameters of aircraft engines, such as thrust, specific fuel consumption, surge margin, etc., can be changed to a certain extent in order to increase aircraft engines function, adaptation and reliability to meet the requirements of different potential users and can be implemented only by using advanced control modes and control laws, causing little change on the aircraft engines' hardware.

The aircraft engine is a complex nonlinear system operated in an uncertain environment of limits: limits in temperature, air pressure and physical acceleration, etc. Aircraft engine control systems can be designed by using linearized engine models under the multiple trimmed flight conditions throughout the flight envelope. For each of these operating points a corresponding linear controller is derived using the well-established linear-based control design methods.3, 4 and 5 However, a problem of this approach is that good performance and robustness properties cannot be guaranteed for a highly nonlinear aircraft engine. Nonlinear control methods have been developed to overcome the shortcomings of linear design approaches.

Several nonlinear control system design methods have emerged over the past two decades.6, 7, 8 and 9 The theoretically established feedback linearization approach is the best known and most widely used among these methods. Feedback linearization is a nonlinear design method that can explicitly handle systems with known nonlinearities. By using nonlinear feedback and exact state transformations rather than linear approximations the nonlinear system is transformed into a constant linear system. This linear system can in principle be controlled by just a single linear controller.10 and 11

During the past decade, switched systems have attracted significant attention, because they can model several practical control problems that involve the integration of supervisory logic-based control schemes and feedback control algorithms. And the results have been developed for linear12 and nonlinear systems.13, 14 and 15 Concerning output feedback control of switched systems, the results are as follows. In Ref.16, by checking the existence of a switched Lyapunov function, linear matrix inequality-based sufficient conditions are derived to deal with the switched static output feedback control of discrete-time switched systems under arbitrary switching. Since there exist linear time-invariant systems, which cannot be stabilized via a single static output feedback, research has been dedicated to the study of hybrid static output feedback stabilization of linear time-invariant systems.[17, 18, 19, 20 and 21] In Ref.[22], the output feedback robust stabilizability problem for uncertain dynamic systems is also considered.

The outline of this paper is as follows. First, the aircraft engine multi-loop control system is described in Section 2 and the control problems are formulated. Section 3 is devoted to the theory of the smooth switching control; a new extended scheme for the smooth switching of a switched control system and an algorithm is also presented in this section. Section 4 shows the procedure to design the aircraft engine multi-loop control system and the simulation results and comparisons. Finally, the conclusion can be obtained in Section 5.

DESCRIPTION OF AIRCRAFT ENGINE MULTI-LOOP CONTROL SYSTEM

Aircraft Engine Descriptions

No matter for the linear or nonlinear control system design methods, the aircraft engine will encounter limits on some occasions. So the aircraft engine control system consists of a family of continuous time subsystems and switches from one to another depending on various environmental factors (see Fig. 1). When the aircraft engine switches from one sub-control loop to another, the stability of the system is the basic performance that must be considered. Therefore, there is much space to improve the dynamic performance of the control system.

In this section, the aircraft engine of two-spool turbofan engine model was developed by the MATLAB simulation environment and its Simulink toolbox. The schematic configuration of the turbofan engine that was simulated is shown in Fig. 2. The high pressure compressor (HPC) and high pressure turbine (HPT) are on one shaft (driven by the high speed rotor), while the fan, booster and low pressure turbine (LPT) are on the other shaft (driven by the low speed rotor). Bleed effects (for air bleed from the booster and the compressor) are not currently considered in the model.

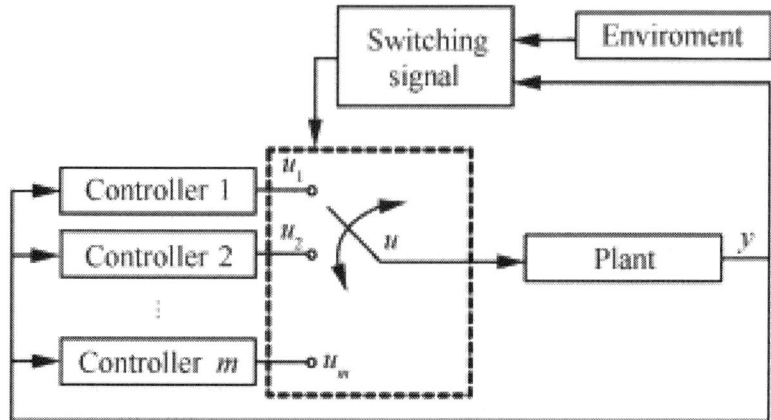

Figure 1: Architecture of multi-controllers.

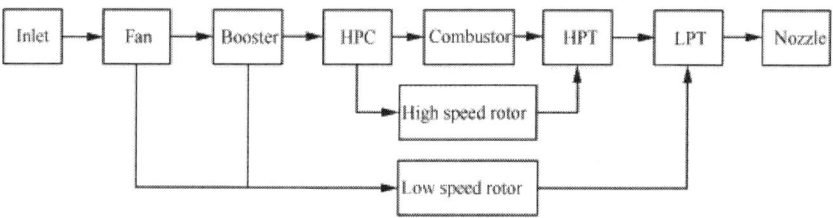

Figure 3: Schematic configuration of two-spool turbofan engine.

The engine simulation model, which is called nonlinear component level (NCL) model, consists of the static elements: inlet, fan (single stage), booster (four stages), high pressure compressor (nine stages), combustor, high pressure turbine (single stage), low pressure turbine (four stages) and main nozzle which are modeled as lumped parameter thermodynamic systems, represented by performance maps, constant coefficients, and thermo and aero-dynamic relationships and the dynamic elements which include low speed rotor and high speed rotor. In the model, the rotor dynamics (for the high speed and low speed rotors) is represented by the equation of conservation of angular momentum. Components, including the fan, the compressor and the turbines, are described in the form of maps and look-up tables based on their individual experimental data. The combustion efficiency and pressure losses are simply fitted by curves. Fig. 3shows the characteristics of the turbofan engine typical components. In Fig. 3, π_F, η_F,$w_{a,cor,F}$ are fan pressure ratio, fan efficiency, and fan corrected air mass flow; π_B, η_B,$w_{a,cor,B}$ are booster pressure ratio, booster efficiency, and booster corrected air mass flow; π_C, η_C, $w_{a,cor,C}$ are compressor pressure ratio, compressor efficiency, and compressor corrected air mass flow.

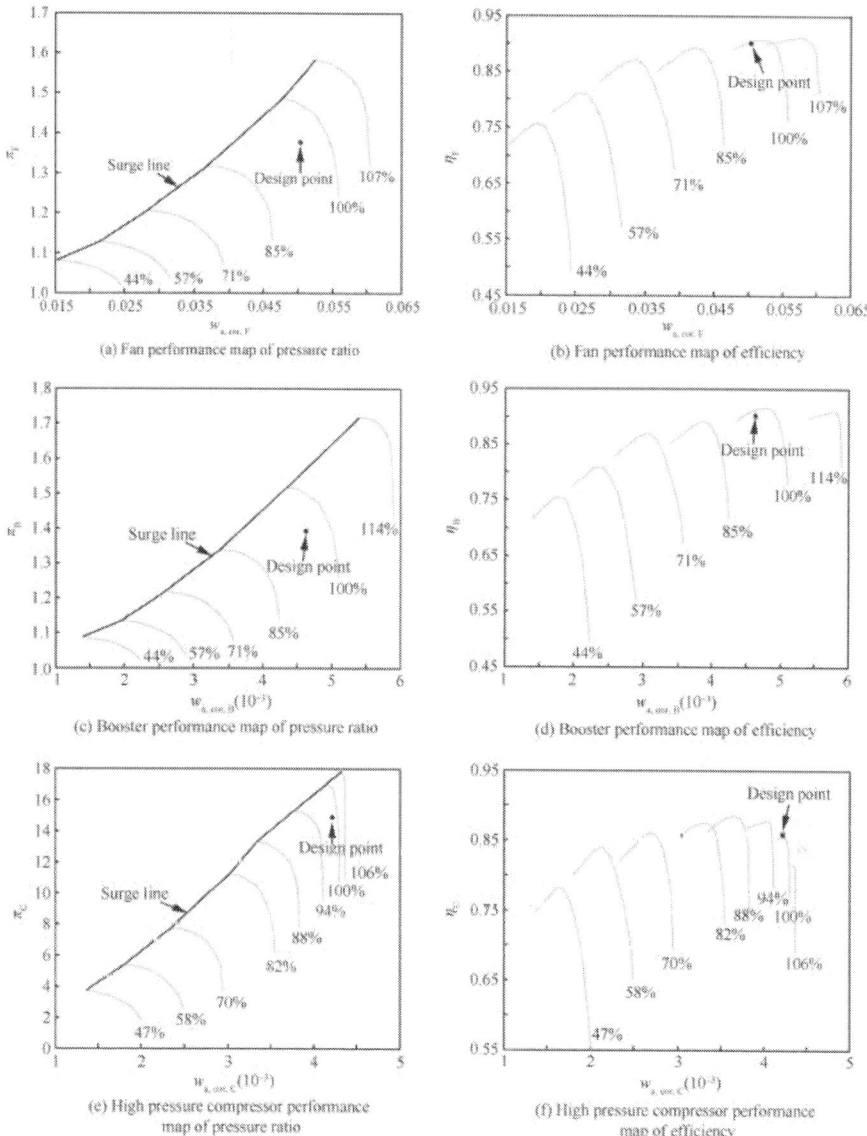

Figure 3: Characteristics of turbofan engine typical components.

Multi-Objective with Aircraft Engine

As a modern aircraft, in order to achieve high performance flights, a high performance aircraft engine control system is essentially indispensable. But there are also many limits during the aircraft engine working process. For example, when the aircraft engine accelerates from one stable condition to another, the compressor surge imposes limits on the aircraft engine operation.

Fig. 4 shows the limit represented by the surge line. The surge line demarcates the regions between stable and unstable operation of the compressor. If the accelerating line goes through the surge line, the stability of the aircraft engine working condition is destroyed. Therefore, the aircraft engine control system has to switch to a stable control loop to avoid the unstable surge oscillation.

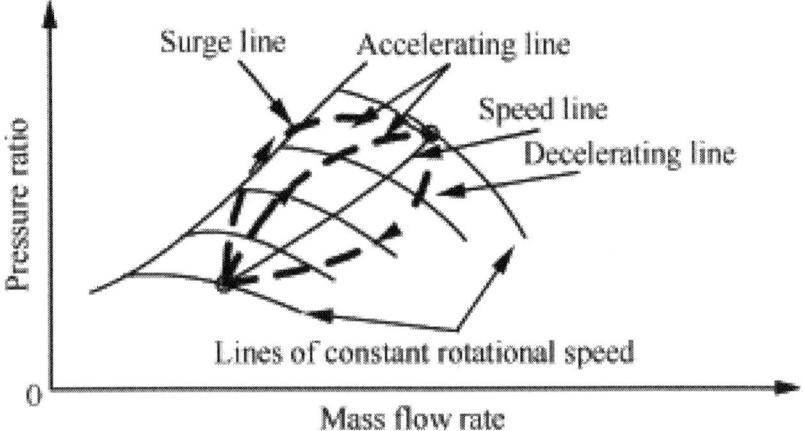

Figure 4: Maps of aircraft engine working characteristics.

Also in the accelerating process, the more fuel flow can cause the excess air coefficient to go through the rich blow-out line (see Fig. 5), resulting in a combustor blow-out event and loss of thrust. So the aircraft engine control system has to switch from the accelerating control loop to another stable control loop, such as speed control loop.

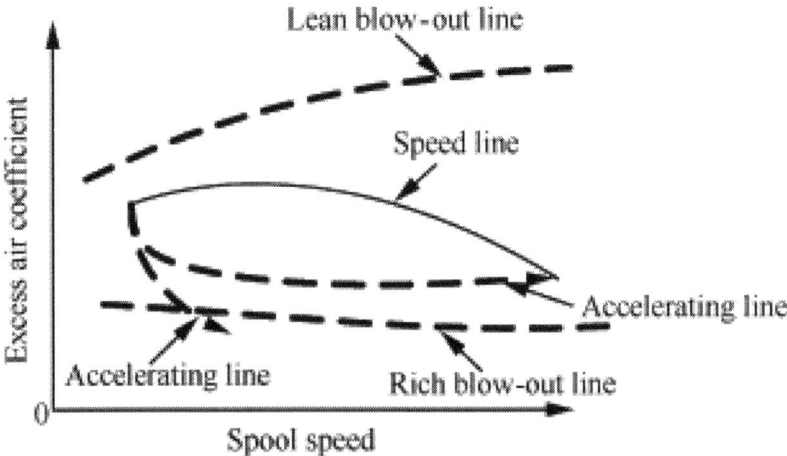

Figure 5: Variation of excess air coefficient with spool speed.

If a limit of turbine temperature is encountered, the remaining command has no effect on an engine. The condition is known as dead-stick and is said to be rather disconcerting.Fig. 6 shows the phenomenon. The pilot lever generates a speed governing and a temperature governing characteristic (Curves 1 and 2, respectively). As speed increases along Curve 1, turbine temperature may rise according to Curve 3. If the temperature reaches its permissible limit before the speed attains its maximum value, Curve 3 must intersect Curve 2. This condition can be detected by equality of the two voltage signals representing a demand and an observed temperature. Control can be smoothly transferred from Curve 1 to Curve 2 over the remaining portion of lever movement, and conversely, as the lever is returned to a lower speed demand. The principle is obviously applicable to several protection controls; whichever parameter reaches its control line first determines the final mode of control. Therefore, the safe working regions are shown in Fig. 7.

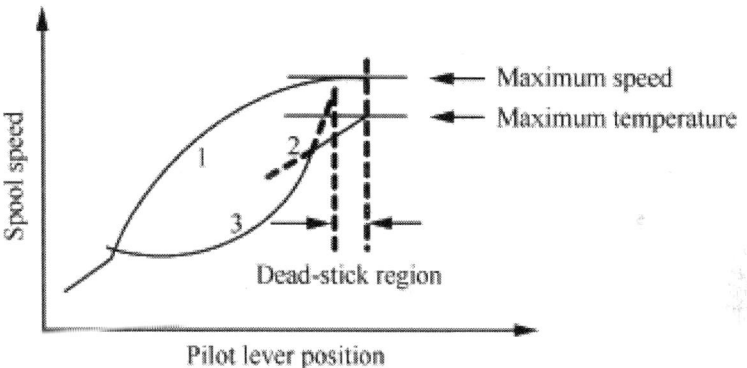

Figure 6: Illustration of dead-stick conditions and possible method for avoidance.

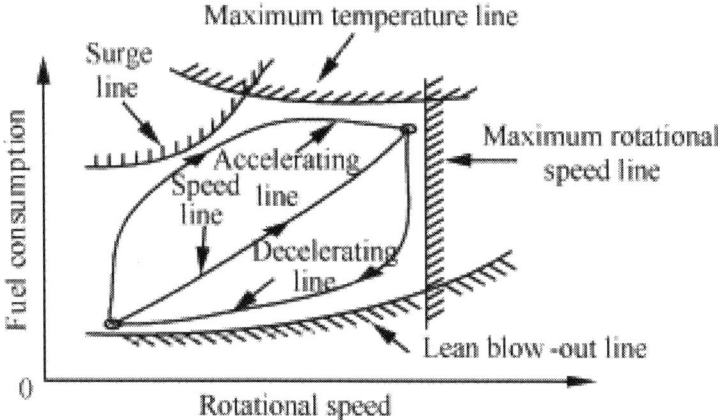

Figure 7: Limits of the aircraft engine works.

Regulation and Protection of Multi-Loops Controllers
As for the rotary speed control of twin shafts engines, one shaft is already controllable, the other one can be controlled by aerodynamic match. Both spools need to be independently protected against over-speed conditions. Fig. 8 illustrates a typical complex control requirement. The speed of the high pressure spool is controlled by the pilot lever angle. To achieve a compromise between engine handling behavior and fuel consumption, the turbine pressure ratio is shown in Fig. 9. When it approaches to maximum low pressure spool speed, turbine pressure ratio is progressively reduced. For further pilot lever angle command, high pressure turbine temperature rises considerably, high pressure spool speed rises slightly, and low pressure spool speed remains almost constant. During this process, in order to prevent the over-speed and over-temperature, the traditional control system makes the aircraft engine's acceleration a slow transition.

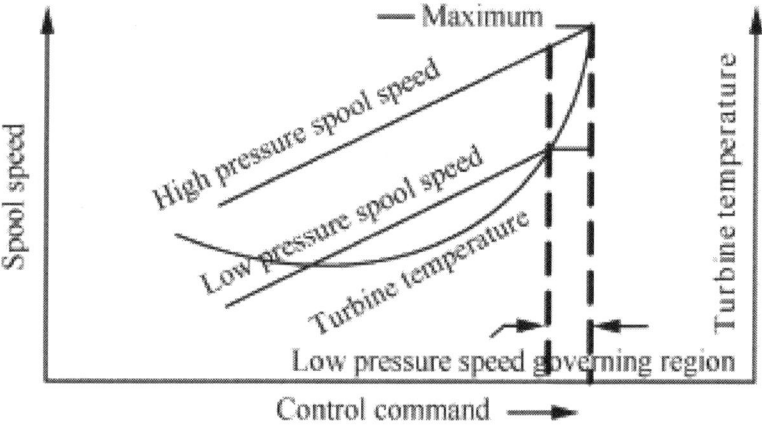

Figure 8: Ilustration of typical engine control requirements.

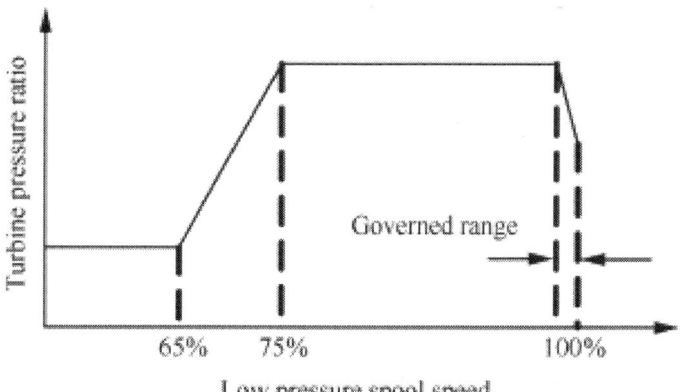

Figure 9: Turbine pressure ratio schedule related to control requirement.

In accelerating control loop, it allows for independent setting of spool speed and turbine temperature, allowing engine working lines to be more closely "fitted" to surge line, with benefit to specific fuel consumption. The speed of response of a nozzle compared with the accelerating time of engine rotor permits rapid thrust control. But the surge margin is an important factor that will influence the accelerating progress, which is shown in Fig. 4.

In the aircraft engine control field, the control requirements of an engine may be classified into the following three aspects: (1) performance controls, for optimum engine operation; (2) limit controls for mechanical integrity; (3) change-of-state controls, mainly for engine handling.

Turbine temperature should not exceed a value determined by the properties of constructional materials. To accelerate an engine, fuel flow is increased; a too rapid increasing-rate of fuel flow may result in a breakdown of the airflow conditions within a compressor, causing engine surge. In order to satisfy the multi-objectives in aircraft engine control, the control system set up includes many sub-control loops, which is shown in Fig. 10. In Fig. 10, n_H is high pressure rotor speed; n_{Href} is reference of n_H; w_f is fuel flow; w_{fref} is reference of w_f and Ref is reference of limit parameter. But during the controller design process, there are many contradictions between the sub-control loops.

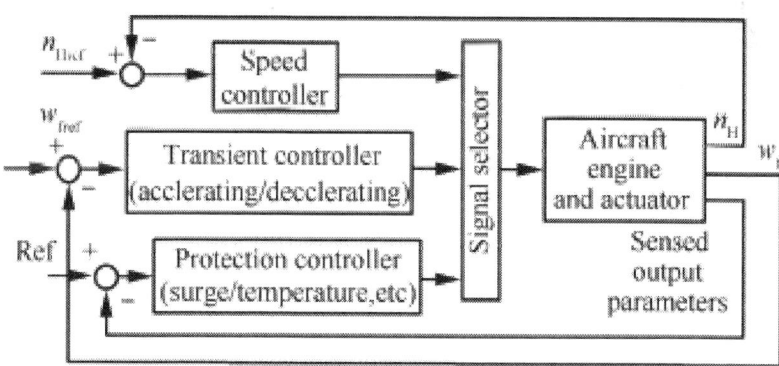

Figure 10: Simplified diagram of the aircraft engine.

Problem of Switching In Aircraft Engine Control

The current application about switching in aircraft engine control is based on Min/Max switching law (see Fig. 11) and focuses mainly on the problem of stability, feedback stabilization. And this switching method is widely applied to engineering practice and is proved effective.

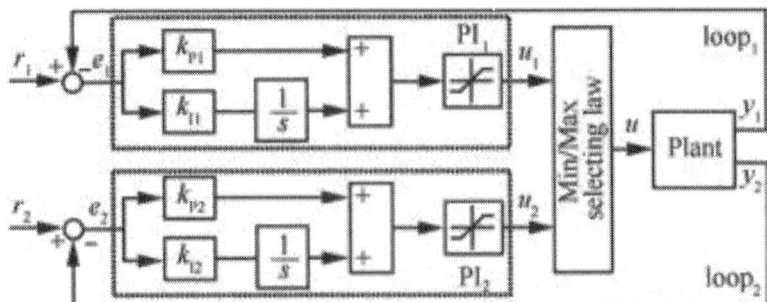

Figure 11: A sketch map for multi-loop switched control system based on Min/Max switching law.

The typical feature of this class of switched control system is that it includes an integral saturating limit and a Min/Max switching law at the same time. The function of integral saturation limiting is to shorten switching delay time so that the system can be switched rapidly; the function of the Min/Max switching law is to select the control signals based on the minimum (maximum) value law. The coupled interactions between the integral saturation limiting and the Min/Max switching law result in that the two controllers are always competing to control the system hence the multi-objective control can be realized.

According to the above Fig. 11, for instance, the control signal u_2 in $loop_2$ is the minimum between the two loops, and $loop_1$ is inactive and will soon reach in a limit state finally due to an integral saturation limiting. When $loop_2$ is motivated under an outside input r_2, the output y_1 may exceed the setting limit r_1 in the control process of y_2, thus the PI_1 must be active to control y_1 in this case. The switching condition is that the control signal of $loop_1$ is equal to and less than that of $loop_2$.

If one controller is at work, the other one will be inactive. After the PI_1 has replaced the PI_2 to take over the system, the accumulated output at the initial stage has been high which may cause an dynamic overshoot of output y_1; moreover, as there is always a non-zero error input to PI_2, high output accumulated after a long integrating period will cause a change in the operation condition, which means it will cost more time to counteract this high output when the system need to be switched from $loop_1$ back to $loop_2$.

According to the shortage of Min/Max switching law, the paper proposes an example to show switching phenomena between the accelerating control loop ($loop_2$) and the temperature limitation control loop ($loop_1$) and the switching strategy is

$$\sigma_{plant}(t) = \begin{cases} 1, & T > T_{max} \\ 2, & T < T_{max} \end{cases} \tag{1}$$

and the simulation is shown in Fig. 12, Table 1, Table 2 and Table 3. In Fig. 12, n_L is low pressure rotor speed; T_4^* is turbine output temperature and Δw_f is variation of fuel flow. In Table 1, Table 2 and Table 3, H is altitude; Ma is Mach number; σ is overshoot and t_s is settle time.

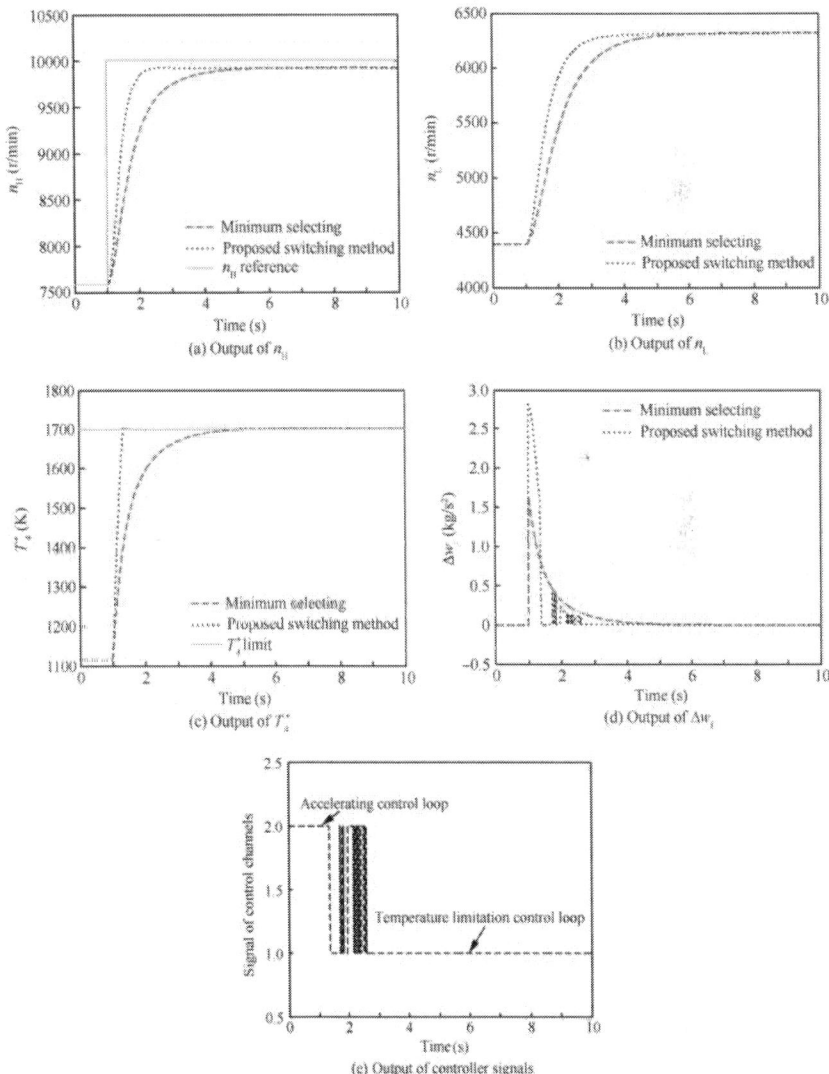

Figure 12: Comparison of two different switching schemes.

Table 1: Difference between two switching schemes ($H = 0$ km, $Ma = 0$).

Performance parameter	σ (%)		t_s (s)	
	Minimum selecting	Proposed switching method	Minimum selecting	Proposed switching method
n_H	0	0.048	1.24	0.61
n_L	0	0	1.9	1.14
T_4^*	0	0.117	1.2	0.29

Table 2: Difference between two switching schemes ($H = 5$ km, $Ma = 0.5$).

Performanc parameter	σ (%)		t_s (s)	
	Minimum selecting	Proposed switching method	Minimum selecting	Proposed switching method
n_H	0	0.048	1.43	0.78
n_L	0	0	2.21	1.48
T_4^*	0	0.359	1.16	0.23

Table 3: Difference between two switching schemes ($H = 8$ km, $Ma = 0.6$).

Performance parameter	σ (%)		t_s (s)	
	Minimum selecting	Proposed switching method	Minimum selecting	Proposed switching method
n_H	0	0.054	1.54	1.04
n_L	0	0	2.59	2.02
T_4^*	0	0.159	0.99	0.32

We assume that the temperature limit is shown in Fig. 12(c). As accelerating sub-control loop fuel flow demand increases, T_4^* approaches the safe limit, and the temperature protection controller engages to control the engine. As expected, there are small glitches during controller switching and the tracking performance is acceptable.

However, this kind of switching method balances the performance and safety, somehow, sacrifices the system performance to guaranty the safety. It can be indicated that a key problem of how to increase switching rapidity to decrease the dynamic response overshoot in switching process should be focused for this class of multi-loop switched control system based on the switching law.

SMOOTH SWITCHING SCHEME DESIGN

Problem Statement and Preliminaries

Consider the switched control system

$$\begin{cases} \dot{x}(t) = Ax(t) + Bu(t), \, x(t_0) = x(0) \\ u(t) = u_{\sigma(t)}(t) \\ y(t) = Cx(t) + Du(t) \end{cases} \tag{1}$$

where $\sigma(t):[0, +\infty) \rightarrow P = \{1, 2, \cdots, N\}$ is the switching signal to be designed; $x(t) \in \mathbf{R}^n$ is the system state, and $u(t) \in \mathbf{R}^m$ the control input; $u_{\sigma(t)}(t) \in \mathbf{R}^m$ represents the sub-controllers outputs and $y(t) \in \mathbf{R}^l$ is the vector of system measurements; A, B, C and D are real constant matrices of appropriate dimensions. Corresponding to the switching signal $\sigma(t)$, we have the switching sequence

$$\Lambda = \{x_0; (i_0, t_0), (i_1, t_1), \cdots, (i_k, t_k), \cdots, | i_k \in P\}$$

When $t \in [t_k, t_{k+1}), \sigma(t) = i_k$, that is, the i_kth subsystem is activated. A set of controllers can be designed in advanced based on the method in Ref.[23] as follows:

$$K_j : \begin{cases} \dot{x}_{K_j}(t) = A_{K_j} x_{K_j}(t) + B_{K_j} \Delta y_i \\ u_i(t) = C_{K_j} x_{K_j}(t) + D_{K_j} \Delta y_i \end{cases} \tag{3}$$

where $x_{Ki}(t) \in \mathbf{R}^{n_K}$ is the i th controller state, and $A_{Ki}, B_{Ki}, C_{Ki}, D_{Ki}$ are the corresponding matrices; Δy_i is the difference between the reference value r_i and the measured system output y_i $(\Delta y_i = r_i - y_i)$.

From the above simulation (shown in Fig. 12), we can find that the modified switching scheme can improve the aircraft engine performance; however, during switching, the switching signal has much oscillation. Therefore, the compensatory controller must be designed to deal this problem.

Integral action is used in classical control to eliminate steady state errors when tracking signals. It can be introduced into the linear quadratic (LQ) framework by considering the integral of the tracking error as an extra set of state variables. For any offline controller κ_i, $i \in P$, define the error state as

$$\dot{\omega}_i = u(t) - u_i(t) = u(t) - (C_{K_j} x_{K_j}(t) + D_{K_j} \Delta y_i) \tag{4}$$

Definition 1 [24]

Given a positive scalar $\varepsilon > 0$, a switching controller κ_i is said to perform a smooth switching if, whenever controller is switched, there exists a finite time $T_\varepsilon > 0$ such that the output of a controller κ_j to be switched satisfies the condition $\lim_{t \to T\varepsilon} |u(t) - u_j(t)| \leqslant \varepsilon$, where $u(t) = u_i(t)$. In particular, κ_i is said to perform a strictly smooth switching $\lim_{t \to \infty} |u(t) - u_j(t)| = 0$.

The offline controller's input is transformed from system reference input r_i into compensatory controller output $r_{C\kappa i}$, Therefore, combining Eqs. (3) and (4), we can get the augmented system for the ith offline controller

$$\begin{bmatrix} \dot{x}_{\kappa_j}(t) \\ \dot{\omega}_i(t) \end{bmatrix} = \begin{bmatrix} A_{\kappa_j} & 0 \\ -C_{\kappa_j} & 0 \end{bmatrix} \begin{bmatrix} x_{\kappa_j}(t) \\ \omega_i(t) \end{bmatrix} + \begin{bmatrix} -B_{\kappa_j} \\ D_{\kappa_j} \end{bmatrix} y_i(t)$$
$$+ \begin{bmatrix} B_{\kappa_j} \\ -D_{\kappa_j} \end{bmatrix} r_{C\kappa_j}(t) + \begin{bmatrix} 0 \\ I \end{bmatrix} u(t) \tag{5}$$

For simplicity, the augmented system (Eq. (5)) is rewritten as

$$\dot{\overline{x}}_i(t) = \overline{A}_i \overline{x}_i(t) + \overline{B}_{i_1} y_i + \overline{B}_{i_2} r_{C\kappa_j} + M u(t) \tag{6}$$

where $\overline{x}_i = \begin{bmatrix} x_{\kappa_j}^T, \omega_i^T \end{bmatrix}^T$ is the augmented system state vector, $\overline{A}_i = \begin{bmatrix} A_{\kappa_j} & 0 \\ -C_{\kappa_j} & 0 \end{bmatrix}$, $\overline{B}_{i_1} = \begin{bmatrix} -B_{\kappa_j} \\ D_{\kappa_j} \end{bmatrix}$, $\overline{B}_{i_2} = \begin{bmatrix} B_{\kappa_j} \\ -D_{\kappa_j} \end{bmatrix}$, $M = \begin{bmatrix} 0 \\ I \end{bmatrix}$.

The compensatory controllers are designed by LQ method. The cost performance index for the design of the ith smooth switching compensatory controller is defined as

$$J_i = \frac{1}{2} \int_0^T \left(\overline{x}_i^T Q_i \overline{x}_i + v_i^T R_i v_i \right) dt \tag{7}$$

where $Q_i \geqslant 0, R_i > 0$ and $v_i = r_{C\kappa i} - r_i$ denote the difference between compensatory controller output $r_{C\kappa i}$ and system's reference signal r_i. Q_i for state variables is standard in optimal control, while the R_i is used for scaling v_i. The size of v_i will seriously influence the error size Eq. (4).

Therefore, when given a set of controllers (such as shown in Eq. (3)), a set of compensatory controllers is designed, respectively, such that the augmented system is stabilized and the number of switching times is minimized. And given a set of controllers and their compensatory controllers, a switching law $\sigma(t)$ is constructed and a sufficient condition is derived such that the closed-loop system composed of Eq. (3), Eq. (3) and compensatory controller is asymptotically stable and simultaneously guarantees the smooth switching by switching controllers.

Compensatory Controller Design

According to the above subsection's description, suppose that there exists a vector function $\lambda_i(t)$, such that the quadratic form performance index function J_i can be rewritten as a non-conditional extremum problem

$$J_i = \frac{1}{2} \int_0^T \left(H_i - \lambda_i^T \dot{\overline{x}}_i \right) dt$$

(8)

where the corresponding Hamiltonian function is

$$H_i = \left(\overline{x}_i^T Q_i \overline{x}_i + v_i^T R_i v_i \right) + \lambda_i^T \left(\overline{A}_i \overline{x}_i + \overline{B}_{i_1} y_i + \overline{B}_{i_2} r_{CK_i} + Mu(t) \right)$$

(9)

The necessary conditions of J_i are

$$\begin{cases} \dfrac{\partial H_i}{\partial \lambda_i} = \dot{\overline{x}}_i \\[2mm] \dfrac{\partial H_i}{\partial \overline{x}_i} = -\dot{\lambda}_i \\[2mm] \dfrac{\partial H_i}{\partial r_{CK_i}} = 0 \end{cases}$$

(10)

Where

$$\frac{\partial H_i}{\partial \overline{x}_i} = Q_i \overline{x}_i + \overline{A}_i^T \lambda_i = -\dot{\lambda}_i$$

(11)

And

$$\frac{\partial H_i}{\partial r_{CK_i}} = R_i (r_{CK_j} - r_i) + \overline{B}_{i_2}^T \lambda_i = 0$$

(12)

And

$$r_{CK_i} = r_i - R_i^{-1} \overline{B}_{i_2}^T \lambda_i$$

(13)

Taking Eq. (13) into Eq. (6), the compensatory controller's state space description is

$$\dot{\overline{x}}_i = \overline{A}_i \overline{x}_i + \overline{B}_{i_1} y_i + \overline{B}_{i_2} \left(r_i - R_i^{-1} \overline{B}_{i_2}^T \lambda_i \right) + Mu(t)$$

(14)

Suppose

$$\lambda_i(t) = P_i(t)\bar{x}_i(t) - g_i(t) \tag{15}$$

And

$$\dot{\lambda}_i(t) = \dot{P}_i(t)\bar{x}_i(t) + P_i(t)\dot{\bar{x}}_i(t) - \dot{g}_i(t) \tag{16}$$

According to Eq. (11)

$$
\begin{aligned}
\dot{\lambda}_i(t) &= -Q_i\bar{x}_i(t) - \overline{A}_i^{\mathrm{T}}(P_i(t)\bar{x}_i(t) - g_i(t)) \\
&= \overline{A}_i^{\mathrm{T}}g_i(t) - (Q_i(t) + \overline{A}_i^{\mathrm{T}}P(t)_i)\bar{x}_i(t)
\end{aligned} \tag{17}
$$

Taking Eqs. (14) and (15) into Eq. (16), we have

$$
\begin{aligned}
\dot{\lambda}_i(t) &= \dot{P}_i(t)\bar{x}_i + P_i(t)(\overline{A}_i\bar{x}_i + \overline{B}_{i_1}y_i + \overline{B}_{i_2}r_i) \\
&\quad - P_i(t)\overline{B}_{i_2}R^{-1}\overline{B}_{i_2}^{\mathrm{T}}(P_i(t)\bar{x}_i - g_i(t)) + P_i(t)Mu - \dot{g}_i(t)
\end{aligned} \tag{18}
$$

Combining Eqs. (17) and (18) yields

$$
\begin{aligned}
&\left(\dot{P}_i + P_i\overline{A}_i + \overline{A}_i^{\mathrm{T}}P_i - P_i\overline{B}_{i_2}R_i^{-1}\overline{B}_{i_2}^{\mathrm{T}}P_i + Q_i\right)\bar{x}_i = \dot{g}_i \\
&\quad + \left(\overline{A}_i^{\mathrm{T}} - P_i\overline{B}_{i_2}R_i^{-1}\overline{B}_{i_2}^{\mathrm{T}}\right)g_i - P\overline{B}_{i_1}y_i - P_i\overline{B}_{i_2}r_i - P_iMu
\end{aligned} \tag{19}
$$

Note that the left side of Eq. (19) is a product of a function of time and state variables $\bar{x}_i(t)$, while the right side is only a function of time. It means that for arbitrary t and $\bar{x}_i(t)$, the following two equations

$$
\begin{cases}
-\dot{P}_i = P_i\overline{A}_i + \overline{A}_i^{\mathrm{T}}P_i - P_i\overline{B}_{i_2}R_i^{-1}\overline{B}_{i_2}^{\mathrm{T}}P_i + Q_i \\
-\dot{g}_i = (\overline{A}_i^{\mathrm{T}} - P_i\overline{B}_{i_2}R_i^{-1}\overline{B}_{i_2}^{\mathrm{T}})g_i - P_i\overline{B}_{i_1}y_i \\
\qquad\quad -P_i\overline{B}_{i_2}r_i - P_iMu
\end{cases} \tag{20}
$$

must be satisfied. When $t \to \infty$, $\dot{P}_i(t) \to 0$ and $\dot{g}_i(t) \to 0$, Eq. (20) can be rewritten as

$$
\begin{cases}
0 = P_i\overline{A}_i + \overline{A}_i^{\mathrm{T}}P_i - P_i\overline{B}_{i_2}R_i^{-1}\overline{B}_{i_2}^{\mathrm{T}}P_i + Q_i \\
0 = (\overline{A}_i^{\mathrm{T}} - P_i\overline{B}_{i_2}R_i^{-1}\overline{B}_{i_2}^{\mathrm{T}})g_i \\
\qquad -P_i\overline{B}_{i_1}y_i - P_i\overline{B}_{i_2}r_i - P_iMu
\end{cases} \tag{21}
$$

And

$$g_i = \left(\overline{A}_i^T - P_i\overline{B}_{i_2}R_i^{-1}\overline{B}_{i_2}^T\right)^{-1}\left(P_i\overline{B}_{i_1}y_i + P_i\overline{B}_{i_2}r_i + P_iMu(t)\right) \quad (22)$$

where P_i is the solution of algebraic Riccati Eq. (21).

Thus, according to Eqs. (13), (15) and (22), the compensatory controller output can be carried out

$$\overline{r}_{CK_j} = K_{r_i}r_i + K_{x_{x_j}}x_{K_j} + K_{\omega_j}\omega_i + K_{y_j}y_i + K_{u_j}u \quad (23)$$

where

$$K_{r_i} = I + R_i^{-1}\overline{B}_{i_2}^T\left(\overline{A}_i^T - P_i\overline{B}_{i_2}R_i^{-1}\overline{B}_{i_2}^T\right)^{-1}P_i\overline{B}_{i_2}, [K_{x_{x_j}} \quad K_{\omega_j}] = -R_i^{-1}\overline{B}_{i_2}^T P_i,$$

$$K_{y_j} = R_i^{-1}\overline{B}_{i_1}^T\left(\overline{A}_i^T - P_i\overline{B}_{i_2}R_i^{-1}\overline{B}_{i_2}^T\right)^{-1}P_i\overline{B}_{i_1}, K_{u_j} = R_i^{-1}\overline{B}_{i_2}^T\left(\overline{A}_i^T - P_i\overline{B}_{i_2}R_i^{-1}\overline{B}_{i_2}^T\right)^{-1}P_iM$$

The design of compensatory controller's output is shown in Fig. 13. An off-line controller is closed by its compensatory controller to drive the control signal close to the online one. And the switching scheme process is shown in Fig. 14.

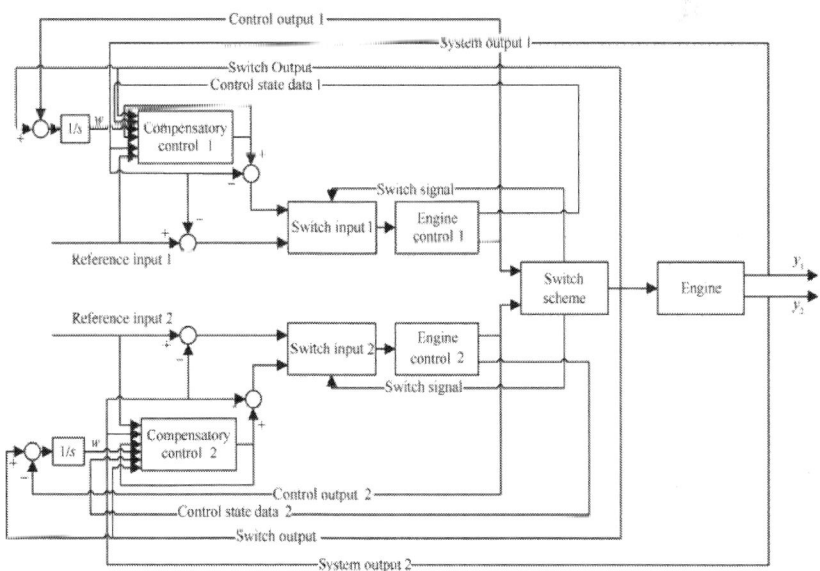

Figure 13: Proposed smooth switching scheme.

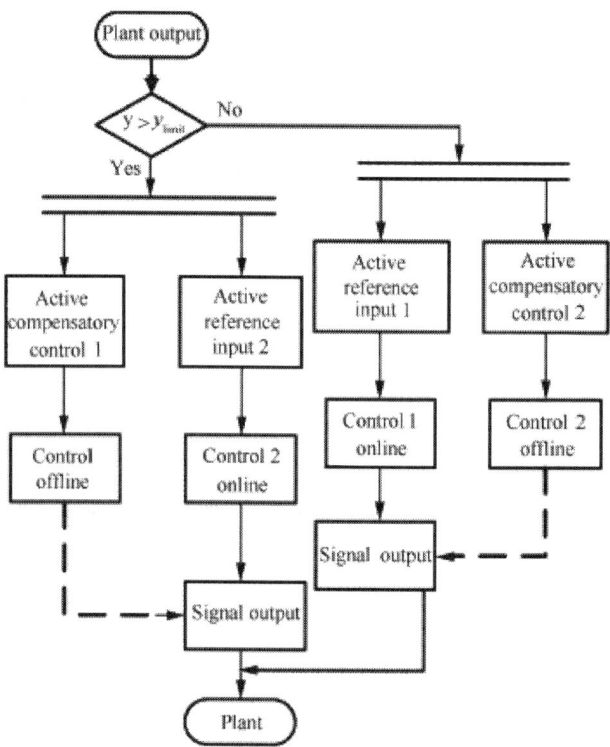

Figure 14: Switching scheme process.

CONTROL DESIGN PROCEDURE AND SIMULATION RESULTS

Augmented Aircraft Engine Model

In order to analyze the time-domain performance objectives, the linear parameter-varying (LPV) model26· 27 [and] 28 of the form is drawn:

$$\begin{cases} \dot{x}(t) = A(\rho(t))x(t) + B(\rho(t))u(t) \\ y(t) = C(\rho(t))x(t) + D(\rho(t))u(t) \end{cases} \tag{24}$$

where $x(t) = [n_H, n_L]^T$, $u(t) = w_f$, the scheduling parameter is selected as $\rho(t) = n_H$, and $y(t\ )$ denotes the aircraft engine measured outputs, such as n_H, n_L, p_3, T_4^*, and performance parameters' outputs, such as F (thrust),

SM$_C$ (surge margin of compressor), and so on. The following simulations show the comparison of LPV model and NCL model.

The results obtained from simulation of the engine n_H and SM$_C$ LPV model are compared with NCL model results in Fig. 15(a). The n_H increases with an increase in w_f. The model output precision is ±1.5%. Throughout the analysis of the operating points, when given the flight condition (such as H = 8 km and Ma = 0.6), the increasing spool speed results in decreasing magnifying coefficient and time coefficient. Furthermore, the dynamic response becomes faster. The comparison of SM$_C$ is shown in Fig. 15(b).

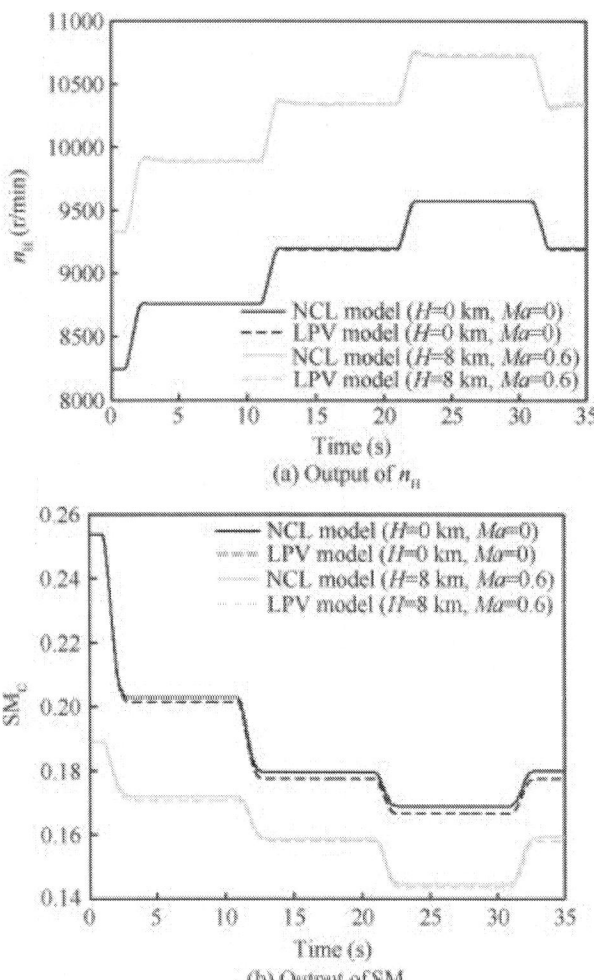

Figure 15:.Comparison of different models at different working points of whole flight envelope.

Smooth Switching Controller Tuning Procedure

Firstly, according to the Section 3.1, the form of accelerating controller, temperature limit controller and surge protection controller is respectively.

$$K_{ACC} : \begin{cases} \dot{x}_{ACC} = A_{ACC}x_{ACC} + B_{ACC}(r_{ACC} - n_{Hmes}) \\ u_{ACC} = C_{ACC}x_{ACC} + D_{ACC}(r_{ACC} - n_{Hmes}) \end{cases} \tag{25}$$

$$K_T : \begin{cases} \dot{x}_T = A_T x_T + B_T(r_T - T^*_{4\,mes}) \\ u_T = C_T x_T + D_T(r_T - T^*_{4\,mes}) \end{cases} \tag{26}$$

$$K_{SMC} : \begin{cases} \dot{x}_{SMC} = A_{SMC}x_{SMC} + B_{SMC}(r_{SMC} - SM_{Cmes}) \\ u_{SMC} = C_{SMC}x_{SMC} + D_{SMC}(r_{SMC} - SM_{Cmes}) \end{cases} \tag{27}$$

where the subscripts of Eqs. (25), (26) and (27) are: "ACC" is accelerating control loop, "T" represents temperature limit control loop, "SMC" is surge margin control loop and "mes" means measured parameter.

The controllers are designed to achieve the following robustness and performance specifications. In the accelerating sub-control loop, the controller is tuned such that steady-state tracking error is zero. The settling time is expected to be within 1.5 s.

In this paper, the mixed sensitivity synthesis design method is used to deal with the three sub-controllers. The sensitivity weighting function $W_S(s)$ is chosen to be large at low frequency in order to obtain good command tracking at low frequency. The complementary sensitivity weighting function $W_T(s)$ is chosen to be large at high frequency to obtain robustness to unmodeled high frequency dynamics. The sensitivity and complimentary sensitivity weighting functions are designed as two parallel first-order lags and leads, respectively, with the low frequency magnitudes and crossover frequencies specified by the user.

Here $W_S(s)$ should have the characteristic of low-pass frequency, and the unity-gain crossover frequencies of 0.4 rad/s. In order to satisfy the bandwidth requirement, $W_T(s)$'s static gain should be adequately small, and it should exhibit a unity-gain crossover frequencies of 100 rad/s. According to the above specifications, the weighting functions are chosen as follows (see Fig. 16).

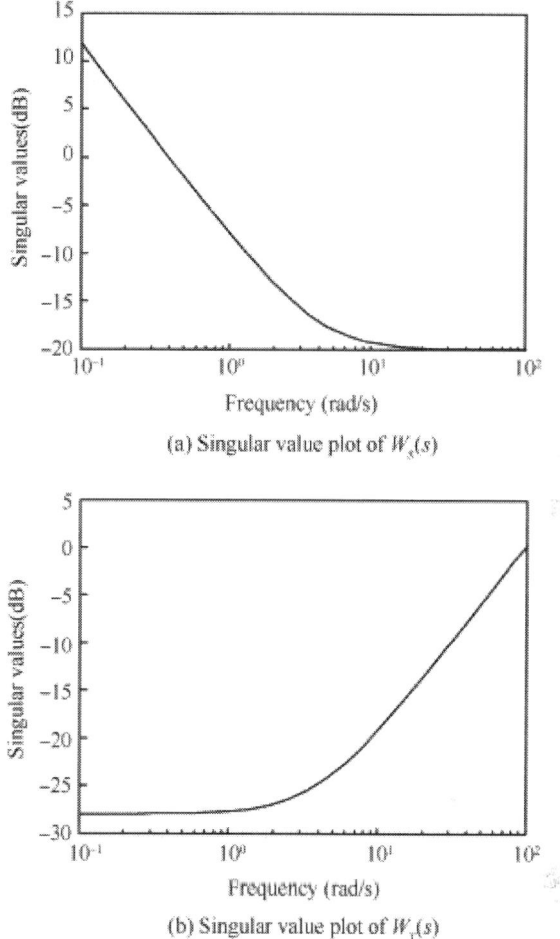

(a) Singular value plot of $W_s(s)$

(b) Singular value plot of $W_t(s)$

Figure 16: Singular value in accelerating control loop.

In the protection/limit sub-control loops, the weighting functions are the same as the accelerating sub-control loop, and the controllers of each control loop are

$$
\begin{cases}
K_{ACC} = \dfrac{0.001s + 10^{-6}}{s} \\[2mm]
K_T = \dfrac{0.005s + 10^{-5}}{s} \\[2mm]
K_{SMC} = \dfrac{-20s + 10^{-5}}{s}
\end{cases}
\tag{28}
$$

Simulation Results

In this section, we use the controllers which are optimized above to the aircraft engine LPV model. The following parameters of accelerating sub-control compensator were designed at $H = 0$ km, $Ma = 0$ operating point and the same condition with temperature limit and surge protection compensators.

$$\overline{A}_{\mathrm{ACC}} = \begin{bmatrix} 0 & 0 \\ 10^{-6} & 0 \end{bmatrix}, \quad \overline{B}_{\mathrm{ACC}_1} = \begin{bmatrix} -1 \\ 0.001 \end{bmatrix},$$

$$\overline{B}_{\mathrm{ACC}_2} = \begin{bmatrix} 1 \\ -0.001 \end{bmatrix}, \quad M = \begin{bmatrix} 0 \\ 1 \end{bmatrix},$$

$$Q = \begin{bmatrix} 11 & 0 \\ 0 & 11 \end{bmatrix}, \quad R = 1,$$

$$P_{\mathrm{ACC}} = \begin{bmatrix} 14.2990301346208 & 10985.7193189240 \\ 10985.7193189236 & 10989035.9518960 \end{bmatrix},$$

$$K_{r_{\mathrm{ACC}}} = -10.0000143166330,$$

$$K_{x_{x_{\mathrm{ACC}}}} = -3.31331081569725,$$

$$K_{\omega_{\mathrm{ACC}}} = 3.31663297194973,$$

$$K_{y_{\mathrm{ACC}}} = 11.0000143166333,$$

$$K_{u_{\mathrm{ACC}}} = 0.011000000029860.$$

$$\overline{A}_{\mathrm{T}} = \begin{bmatrix} 0 & 0 \\ 10^{-5} & 0 \end{bmatrix}, \quad \overline{B}_{\mathrm{T}_1} = \begin{bmatrix} -1 \\ 0.005 \end{bmatrix},$$

$$\overline{B}_{\mathrm{T}_2} = \begin{bmatrix} 1 \\ -0.005 \end{bmatrix}, \quad M = \begin{bmatrix} 0 \\ 1 \end{bmatrix},$$

$$Q = \begin{bmatrix} 11 & 0 \\ 0 & 11 \end{bmatrix}, \quad R = 1,$$

$$P_{\mathrm{T}} = \begin{bmatrix} 30.6464335407947 & 5469.26808301907 \\ 5469.26808301941 & 1094516.94148650 \end{bmatrix},$$

$$K_{r_{\mathrm{T}}} = -10.0003081662441,$$

$$K_{x_{x_{\mathrm{T}}}} = -3.30009312569769,$$

$$K_{\omega_{\mathrm{T}}} = 3.31662441345543,$$

$$K_{y_{\mathrm{T}}} = 11.0003081662441,$$

$$K_{u_{\mathrm{T}}} = 0.0549999999977263.$$

Simulation 1: Switching process between accelerating and temperature sub-loops. In order to show the switching between the two sub-controllers in a harsh condition, the reference input of the accelerating speed is given in Fig. 17(a).

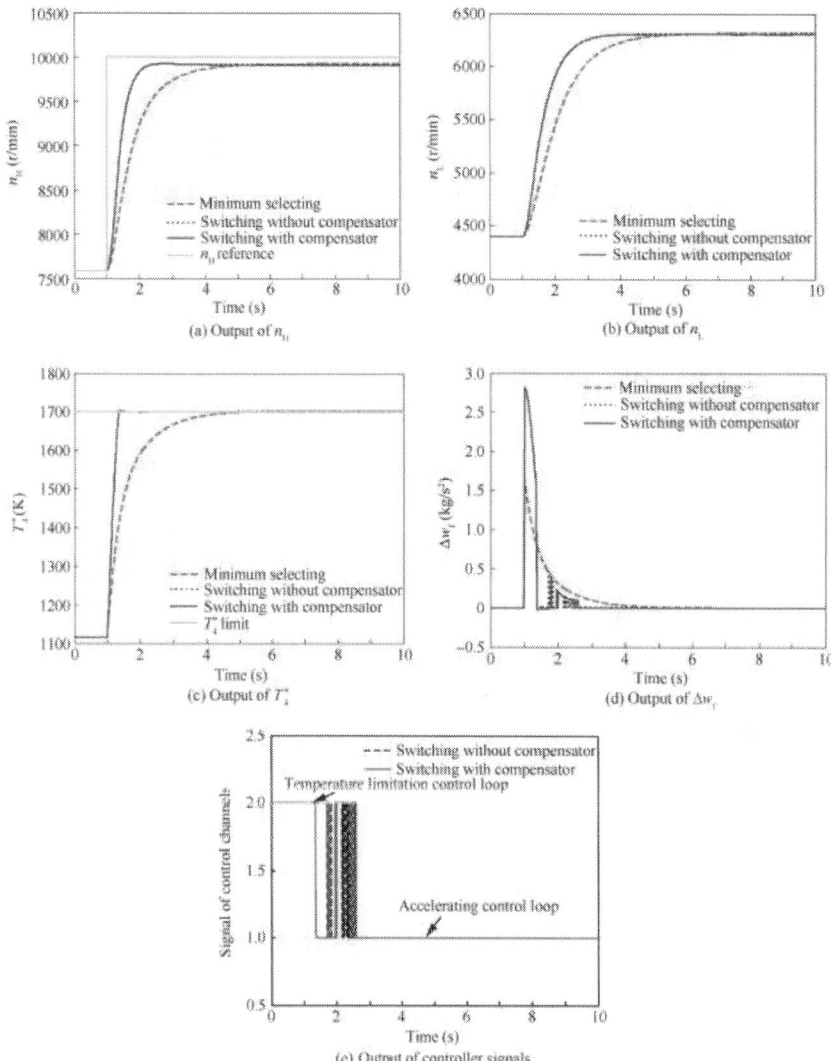

Figure 17: Comparison of different switching schemes (accelerating switching with temperature limitation sub-control loops at $H = 0$ km, $Ma = 0$).

Assume that the temperature limit is as shown in Fig. 17(c). As accelerating sub-control loop speed demand increases, the temperature approaches the safe limit, and the temperature limit controller is in charge of controlling the engine. As expected, there are small glitches during controller switching, and the tracking performance is acceptable. The difference of performance between two switching schemes is shown in Table 4.

Table 4: Difference between two switching schemes (accelerating switching with temperature limitation sub-control loops at $H = 0$ km, $Ma = 0$).

Perfor-mance para-meter	σ (%)			ts (s)		
	Minimum selecting	Switching without compensator	Switching with compensator	Minimum selecting	Switching without compensator	Switching with compensator
n_H	0	0.048	0.134	1.24	0.61	0.6
n_L	0	0	0	1.9	1.14	1.13
T_4^*	0	0.117	0.241	1.2	0.29	0.28

Simulation 2: Switching process between accelerating and surge protection two sub-loops.

$$\overline{A}_{SMC} = \begin{bmatrix} 0 & 0 \\ 10^{-5} & 0 \end{bmatrix}, \quad \overline{B}_{SMC_1} = \begin{bmatrix} -1 \\ -20 \end{bmatrix},$$

$$\overline{B}_{SMC_2} = \begin{bmatrix} 1 \\ 20 \end{bmatrix}, \quad M = \begin{bmatrix} 0 \\ 1 \end{bmatrix},$$

$$Q = \begin{bmatrix} 0.05 & 0 \\ 0 & 0.05 \end{bmatrix}, \quad R = 1,$$

$$P_{SMC} = \begin{bmatrix} 79419491.887 & -3970974.150 \\ -3970974.150 & 198548.696 \end{bmatrix},$$

$$K_{r_{SMC}} = -19.0500022005290,$$

$$K_{x_{NSMC}} = -8.87945564091206,$$

$$K_{\omega_{SMC}} = 0.220086625311524,$$

$$K_{y_{SMC}} = 20.0500022005290,$$

$$K_{u_{SMC}} = -0.999999993946403.$$

The simulation results are shown in Fig. 18. We can obtain the detailed characterization as shown in Table 5, the smooth switching performance indices of overshoot and settling time of SM$_C$ obtained by Min selecting law and switching without compensator are 5.878%, 10.8%, and 2.19 s, 1.73 s, respectively. However, the indices obtained by switching method with compensator are much smaller, which are 0.241% and 1.71 s, respectively. Thus, it clearly demonstrates that the proposed method achieves a better smooth switching performance.

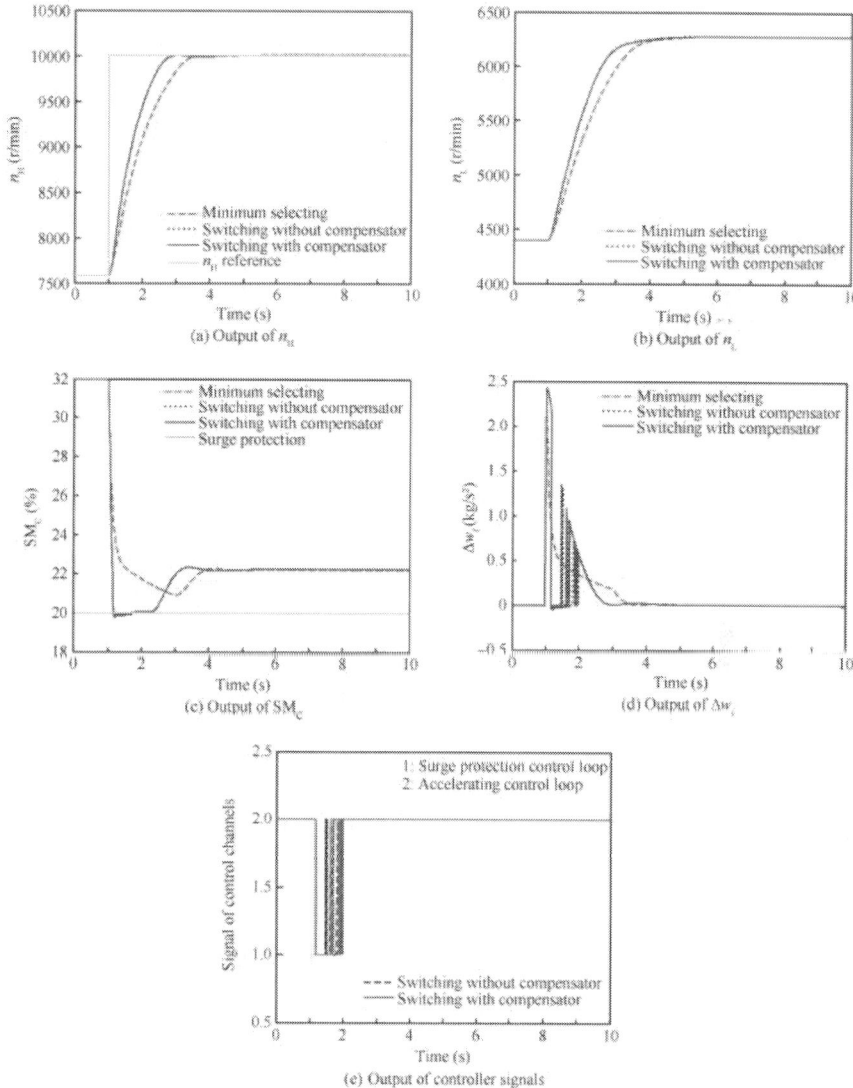

Figure 18: Comparison of different switching schemes (accelerating switching with surge protection sub-control loops at $H = 0$ km, $Ma = 0$).

Table 5: Difference between two switching schemes (accelerating switching with surge protection sub-control loops at $H = 0$ km, $Ma = 0$).

Performance parameter	σ (%)			ts (s)		
	Minimum selecting	Switching without compensator	Switching with compensator	Minimum selecting	Switching without compensator	Switching with compensator
n_H	0	0	0	1.47	1.05	1.05
n_L	0	0	0	1.96	1.51	1.51
SM_C	5.878	10.8	0.241	2.19	1.73	1.71

CONCLUSIONS

In this paper, the aircraft engine multi-loop control system is described in detail and the switching control theory is introduced to solve the regulating and protecting control problems. This paper might give a positive thought that the switching performance objectives and the switching scheme are given and a family of PID controllers and compensators is designed.

Firstly, a new switching scheme has been proposed, throughout the comparison with traditional switching method (Min/Max switching law), the performance indices can be improved better. However, the frequency switching cannot be avoided. After that, a smooth switching method has been investigated based on the optimal theory and the internal model principle. Finally, two examples have confirmed the effectiveness of the proposed bumpless switching approach. The simulations show that using the method can not only improve the dynamic performance of the engine control system, but also can guarantee the stability in some occasions. In that way, the conservatism is reduced, the dynamic performance is improved and the multi-loop controllers are regulated well.

ACKNOWLEDGMENTS

The authors are grateful to Prof. Ming Cao and Dr. Tao Liu from University of Groningen for discussions. They would also like to thank the anonymous reviewers for their critical and constructive review of the manuscript. This study was supported by the National Natural Science Foundation of China (Grant No. 61104146/F030203) and Innovation Plan of Aero Engine Complex System Safety by the Ministry of Education Chang Jiang Scholars of China (Grant No. IRT0905).

REFERENCES

1. Skira CA, Agnello M. Control systems for the next century's fighter engines. J Eng Gas Turbines Power 1992;114(4):749–54.
2. Jaw LC, Mattingly JD. Aircraft engine controls: design, system analysis and health monitoring. Reston: AIAA; 2009.
3. Blakelock JH. Automatic control of aircraft and missiles. New York: Wiley; 1991.
4. Garg S. Robust integrated flight/propulsion control design for a STOVL aircraft using H1 control design techniques. Automatica 1993;29(1):129–45.
5. Liu XF, Zhao L. Approximate nonlinear modeling of aircraft engine surge margin based on equilibrium manifold expansion. Chin J Aeronaut 2012;25(5):663–74.
6. Isidori A. Nonlinear control systems. New York: Springer-Verlag; 1999.
7. Littleboy DM, Smith PR. Using bifurcation methods to aid nonlinear dynamic inversion control law design. J Guid Control Dyn 1998;21(4):632–8.
8. Snell SA, Nns DF, Arrard WL. Nonlinear inversion flight control for a supermaneuverable aircraft. J Guid Control Dyn 1992;15(4): 976–84.
9. Mulgund SS, Stengel RF. Aircraft flight control in wind shear using sequential dynamic inversion. J Guid Control Dyn 1995; 18(5):1084–91.
10. Samar R, Postlethwaite I. Multivariable controller design for a high performance aero-engine. Proceedings of international conference on control, Vol. 94; 1994. p. 1312 7.
11. Fredcrick DK, Garg S, Adibhatla S. Turbofan engine control design using robust multivariable control technologies. IEEE Trans Control Syst Technol 2000;8(6):961–70.
12. Peleties P, Decarlo R. Asymptotic stability of m-switched systems using Lyapunov-like functions. Proceedings of American control conference; 1991 Jun 26–28; Boston, MA; 1991. p. 1679–84.
13. Branicky MS. Multiple Lyapunov functions and other analysis tools for switched and hybrid systems. IEEE Trans Autom Control 1998;43(4):475–82.
14. Pettersson S, Lennartson B. Stability and robustness for hybrid systems. Proceedings of the 35th IEEE conference on decision and control; 1996 Dec 11–13; Kobe. 1996. p. 1202–7.
15. Ye H, Michel AN, Hou L. Stability theory for hybrid dynamical systems. IEEE Trans Autom Control 1998;43(4):461–74.

16. Daafouz J, Riedinger P, Iung C. Stability analysis and control synthesis for switched systems: a switched Lyapunov function approach. IEEE Trans Autom Control 2002;47(11):1883–7.

17. Hu B, Zhai G, Michel A. Hybrid static output feedback stabilization of second-order linear time-invariant systems. Linear Algebra Appl 2002;351:475–85.

18. Liberzon D. Stabilizing a linear system with finite-state hybrid output feedback. Proceedings of the 7th IEEE mediterranean conference on control and automation; 1999.

19. Litsyn E, Nepomnyashchikh YV, Ponosov A. Stabilization of linear differential systems via hybrid feedback controls. SIAM J Control Optim 2000;38(5):1468–80.

20. Artstein Z. Examples of stabilization with hybrid feedback. Hybrid systems III. Heidelberg: Springer; 1996. p. 173–85.

21. Zhai G, Chen X. Stabilizing linear time-invariant systems with finite-state hybrid static output feedback. Proceedings of IEEE international symposium on circuits and systems; 2002. p. 249– 52.

22. Savkin AV, Evans RJ. Robust output feedback stabilizability via controller switching. Proceedings of the 35th IEEE conference on decision and control; 1996 Dec 11–13; Kobe. 1996. p. 2654–8.

23. Mu J, Rees D, Liu GP. Advanced controller design for aircraft gas turbine engines. Control Eng Pract 2005;13(8):1001–15.

24. Bao W, Qi Y, Yu D, Yao Z, Zhao J. Bumpless switching scheme design and its application to hypersonic vehicle model. Int J Innov Comput Inf Control 2012;8(1):677–89.

25. Sternad M, Ahle´n A. LQ controller design and self-tuning control. IEE Control Eng Ser 1993;56.

26. Bruzelius F. Linear parameter-varying systems dissertation. Go¨teborg, Sweden: Chalmers University of Technology; 2004.

27. Balas GJ. Linear, parameter-varying control and its application to a turbofan engine. Int J Robust Nonlinear Control 2002;12(9): 763–96.

28. Bamieh B, Giarre L. Identification of linear parameter varying models. Int J Robust Nonlinear Control 2002;12(9):841–53

CITATION

Xiaofeng Liu, Jing Shi, Yiwen Qi, Ye Yuan, Design for aircraft engine multi-objective controllers with switching characteristics, Chinese Journal of Aeronautics, Volume 27, Issue 5, October 2014, Pages 1097-1110, ISSN 1000-9361, http://dx.doi.org/10.1016/j.cja.2014.08.002

CHAPTER 5

Design, Manufacturing and Test of a High Lift Secondary Flight Control Surface with Shape Memory Alloy Post-Buckled Precompressed Actuators

Thomas Sinn[1] and Ron Barrett[2],

[1]Mechanical & Aerospace Engineering Department, University of Strathclyde, Glasgow G1 1XJ, UK; E-Mail: thomas.sinn@strath.ac.uk
[2]Aerospace Engineering Department, the University of Kansas, Lawrence, KS 66045, USA

ABSTRACT

The use of morphing components on aerospace structures can greatly increase the versatility of an aircraft. This paper presents the design, manufacturing and testing of a new kind of adaptive airfoil with actuation through Shape Memory Alloys (SMA). The developed adaptive flap system makes use of a novel actuator that employs SMA wires in an antagonistic arrangement with a Post-Buckled Precompressed (PBP) mechanism. SMA actuators are usually used in an antagonistic arrangement or are arranged to move structural components with linearly varying resistance levels similar to springs. Unfortunately, most of this strain energy is spent doing work on the passive structure rather than performing the task at hand, like moving a flight control surface or resisting air loads. A solution is the use of Post-Buckled Precompressed (PBP) actuators that are arranged so that the active elements do not waste energy fighting passive structural stiffnesses. One major problem with PBP actuators is that the low tensile strength of the piezoelectric elements can often result in tensile failure of the actuator on the convex face. A solution to this problem is the use of SMA as actuator material due to their tolerance of tensile stresses. The power consumption to hold deflections is reduced by approximately 20% with the Post-Buckled Precompressed mechanism. Conventional SMAs are essentially non-starters for many classes of aircraft due to the requirement of holding the flight control surfaces in a given position for extremely long times to trim the vehicle. For the reason that PBP actuators balance out air and structural loads, the steady-state load on the SMAs is essentially negligible, when properly designed. Simulations and experiments showed that the SMAPBP actuator shows tip rotations on the order of 45°, which is nearly triple the levels achieved by piezoelectric PBP actuators. The developed SMAPBP actuator was integrated in a NACA0012 airfoil with a flexible skin to carry out wind tunnel tests.

INTRODUCTION

For more than 20 years, adaptive materials have been regularly integrated into aerospace vehicles, structures and surfaces of many sizes. From F-14s to Boeing 787s their ability to change physical properties or state on demand has made them desirable for many different devices. Starting with simple bending- and twist-active plates in the late 1980s, some of the first rudimentary flap-type devices were seen as early as 1990 [1,2,3,4]. The first attempts at making adaptive rotors, missile fins, wings and stabilizers were also publicly exhibited in 1990 with the first patent on such devices issuing in 1995 [5,6,7,8]. These devices showed great promise and lead to numerous aeroelastic studies and development of missile fins which were capable of nontrivial deflection levels, high rates and very compact form factors [9,10,11,12,13,14,15,16,17,18]. Several early studies explored the potential for using adaptive materials on hard-launched munitions and air to ground weapon systems [19,20,21,22,23,24,25,26,27,28,29]. A number of helicopter rotor studies drawing their lineage back to the late 1980s and continuing through the 1990s showed fundamental utility of the piezoelectric materials as rotor blade actuators [6,7,30,31,32]. These many advances were shown to be "enabling" for some of the smallest aircraft commissioned by the Department of Defense (DoD). As part of the DoD CounterDrug Technology Office's remote sensing program, the DoD's first Micro Aerial Vehicle (MAV) program kicked off in 1994 and concluded with hundreds of successful flights by 1997. High bandwidths (>40 Hz), large deflections (>±10°) and extremely low mass (<400 mg) requirements made for a challenging program [33,34]. This lead to DARPA's first MAV program which pioneered in developing both fixed and rotary-wing subscale aircraft.

Private funding continued the lineage of adaptive aircraft and resulted in the first ultra-high performance, convertible UAVs. Capable of hovering in more places than a helicopter, then dashing out at missile speeds, through 350 kts in a standard atmosphere with more than an hour of dash capability, the XQ-138 possessed performance over a decade ago that was not matched by any aircraft of its class that was sponsored by the US Federal government. As part of this study, several high performance kits were developed. In an effort to drive component weights lower with higher performance, new techniques were sought to boost piezoelectric actuator mechanical energy and power densities. One paradigm-shifting advance in this area was made in 1997 when it was discovered that with proper design, piezoelectric elements could possess effective coupling coefficients (and therefore electrical-to-mechanical conversion efficiencies) greater than that of the basic material [35,36]. Although these basic advances were applied only to electrical transformers, their implications for flight control actuation was clear, and resulted in a number of high performance actuators. These Post-Buckled Precompressed (PBP) actuators possessed nearly four times greater deflection levels than their linear counterparts while maintaining all of the force and moment generation capability [37,38,39,40,41]. These actuators were integrated into speed kits for the XQ-138 aircraft which boosted flight speeds by nearly 30% while reducing flight control weight fractions by a factor of 4.8 with respect to conventional electromagnetic flight control actuators.

Although highly successful, piezoelectrically energized PBP actuators possess some profound limitations. The techniques pioneered in References [37,42] showed that through dynamic elastic axis shifting and proper control, extremely high speeds and robust performance can be achieved; however, the elements are still limited.

Following the commercial serial production of XQ-138 PBP actuators, a host of other efforts proliferated. Many of these efforts, like the concept of post-buckled precompressed actuators themselves drew their roots back to the original work of Lesieutre in the late 1990s [35,36]. This, of course, is critical to note, as it was discovered back then that the amount of mechanical strain energy and power consumed could be almost completely nulled by the application of suitable axial force levels. While the team spearheading the XQ-138 program chose to operate just below buckling levels, other groups operated just above them, leading to a host of bi-stable devices. Arrieta, Bilgen, Friswell and Hagedorn employed bi-stable morphing concepts for passive load alleviation [43]. The Friswell team continued to explore many different arrangements of flight control mechanisms including inextensible airfoils [44]. Energy harvesting was enabled by using near-buckling and bi-stable structures fitted with piezoelectric elements [45,46,47,48,49,50,51]. One general property that was observed in many of these studies was that with greater levels of near-buckling deflection amplification levels, hysteresis levels were increased dramatically. While the XQ-138 team simply sensed position and closed a loop to solve the problem, high end structural-mechanics models were developed to attack the open-loop variant of the problem. These general concepts were also extended to including hysteresis models and into devices using single crystal piezoelectric transducers [52,53,54]. Because piezoelectric macro-fiber composites could endure relatively high strain levels, they were employed in a number of flight control surfaces, some of which made it into flight quite successfully [55,56,57,58,59]. To enable these devices a nontrivial amount of work was done on the driving electronics (especially reducing them to flightweight levels) [60].

With limited tensile stress levels to fracture, piezoelectrically driven PBP actuators rapidly can generate so much deflection that they literally tear themselves apart. Accordingly, this paper is centered on a change of actuator material class which allows the benefits of PBP actuation without the hazard of self-destruction. By using Shape Memory Alloys (SMA) in a PBP configuration, some profound benefits are realized. The first is that even at extremely high deflection levels, self-destruction simply does not occur. The second aids with a fundamental problem possessed by PBP actuators: They are fundamentally inefficient electrical-to-mechanical transducer mechanisms. With a PBP actuator configuration, much of this issue is overcome as efficiency levels are boosted often by an order of magnitude. Finally, the power consumption to hold deflection is reduced by one if not two orders of magnitude. Because aircraft often require flight control surfaces to be held in a given position for extremely long times to trim the vehicle, conventional SMAs are essentially non-starters for many classes of aircraft. This is mainly due to their speed and dominant irreversible energy loss. Because PBP actuators balance out air and structural loads, the steady-state load on the SMAs is essentially negligible, when properly designed. These advantages of SMAPBP actuators will be explored along with basic structural mechanics, proof-of-concept experiments and correlation.

SMAPBP ACTUATOR

The principal design of the actuator was a plate with constant thickness but tapered from root to tip to account for air loads. The actuator plate itself was a composite plate with spring-steel fiberglass epoxy layers. A prove of concept study for this kind of SMA actuated PBP actuator was undertaken in reference [43]. The actuators were designed to function with an applied axial buckling load close to the perfect column buckling load of the actuator itself. The test specimen consisted of a tapered composite plate with a spring steel substrate and two 45° glass fiber epoxy composite layers on both sides. The actuator had an overall length of 135 mm with 90 mm of active structure that could be actuated by the SMA wire. Due to the fact that air loads result in high root stresses, the plate itself was tapered from 30 mm at the root to 10 mm at the tip. Preliminary investigations showed that the spring steel substrate available in the lab with a stiffness of 207 GPa and a thickness of 0.127 mm was appropriate to use in the selected design. To actuate the steel composite plate, SMA filaments were attached to the top and bottom of the composite plate. An SMA wire with a 0.254-millimeter diameter, with a stiffness of 75 GPa in the austenitic and 28 GPa in the martensitic state, was selected. An SMA actuation wire of 0.254 millimeter diameter Dynalloy Flexinol [60] was used in these experiments. Prior to the assembly of the actuator, the shape memory alloy wire was prestrained to 2500 µs train plastic strain. The glass fiber layers had the only purpose to create a nonconductive layer between the spring steel substrate and the conductive SMA wire. The 45° orientation was chosen to minimize the impact of the glass fiber layer on the overall stiffness of the plate. The glass fiber cloth layers had a stiffness of 4.519 GPa and a thickness of 0.222 mm for each of the four glass fiber layers.

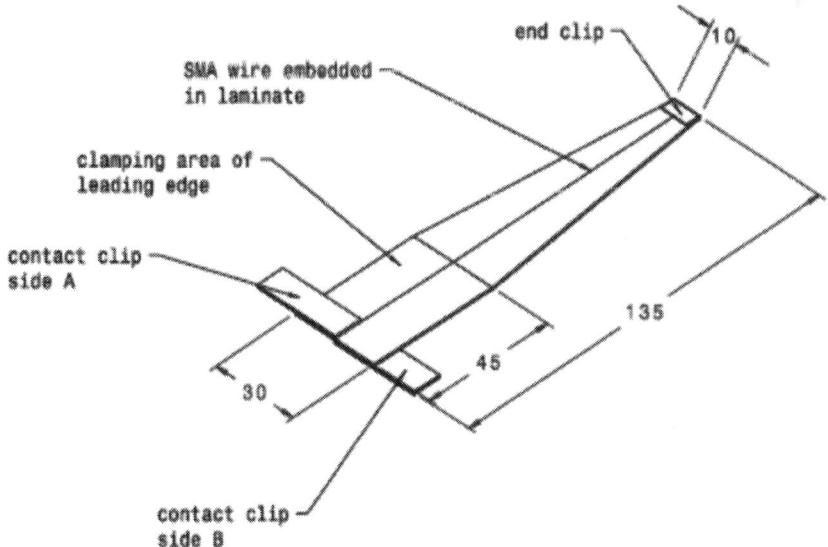

Figure 1: SMAPBP actuator.

The atomic structure changes from Martensite to Austenite by applying heat to the SMA wire. This atomic structure change results in a contraction of the SMA wire and the plate starts to bend. Movement of the SMA wire was suppressed by a clamped steel plate (tip clip) at the tip and a welded steel clip (contact clip) at the root of the test specimen. One single SMA wire was used for actuation of both sides; it was threaded through a transfixion at the tip of the SMAPBP specimen and fixed by a clamped steel plate close to the root of the specimen, the contact clips. The steel substrate served as an electrical conductor in order to produce a closed electrical circle. This arrangement made it possible to actuate both sides separately. For actuation of one side electric power was applied between one of the contact clips and the steel substrate. The steel substrate and the SMA wire were conductively connected by the tip clip (Figure 1).

Due to the observations made in former actuator design steps, it was a natural progression to place the SMA on top of the composite substrate. To prevent the SMA wire from interfering with the bond line between actuator and leading edge, the SMA wire was integrated into the composite at the leading edge. The SMA wire was embedded between two layers of glass fiber epoxy cloth at the clamping area. The 90 mm active length gave the airfoil with integrated actuator an overall length of 150 mm with a 60 mm long leading edge.

Experimental Set-Up

The test specimen was clamped into the test apparatus in a vertical position. To simulate the axial buckling force F_a, an elastic band with weights was attached to the tip of the test specimen. The band ran along the whole length of the specimen, weights could easily be added under the test specimen (see Figure 2).

The specimen was loaded in 100 g steps which led to axial force steps of approximately 1 N. A mirror at the tip of the actuator was used to measure the tip rotation of the specimen. A laser beam from a light source leveled to the height of the tip was reflected to a displacement measuring device in the distance (Figure 2). This test setup made it possible to measure the tip rotation angle accurately. By applying an electric current on either the top or the bottom SMA wire, the SMA wire started to heat up and therefore contract, the SMAPBP actuator started to bend.

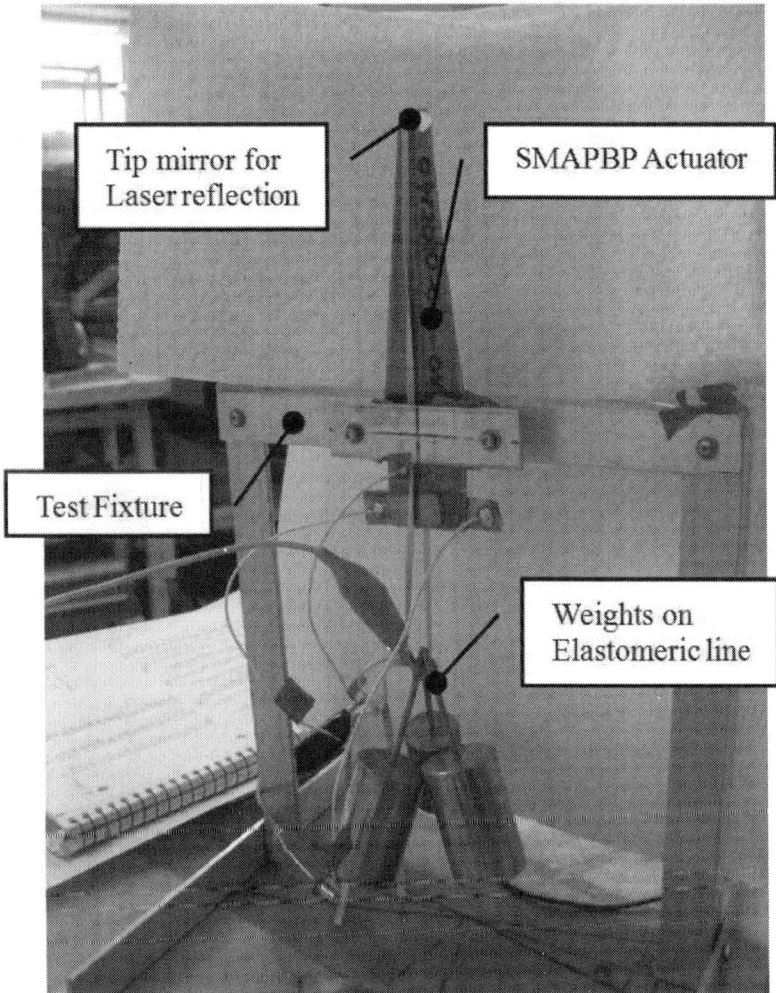

Figure 2: SMAPBP actuator clamped in test fixture to measure tip rotation.

Results

Figure 3 clearly shows a trend to higher tip rotation angles by increasing the applied axial load. The figure also shows that the displacement with no applied current increases with increasing axial force, this may be caused by bending and buckling phenomena from the applied axial load. The slope of the curves for the different load cases followed also the same pattern. The actuator had nearly no response in the first part of the slope (0 A < I < 0.4 A), followed by a steep rise (0.4 A < I < 0.75 A) and a plateau phase where the tip rotation angle seems to reach a maximum in the order of 55° to 70° dependent on the applied buckling load.

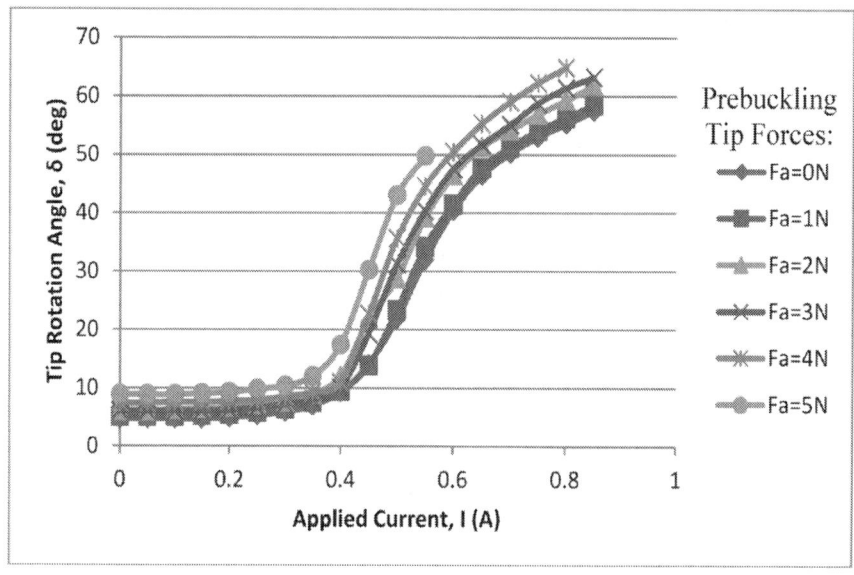

Figure 3: Experimental data: tip rotation angle over applied current.

Figure 3 shows the experimental data for the tip rotation over the applied buckling force F_a for the non-actuated case and the actuated case. The figure shows that the test specimen already had a tip rotation of 4.76° at the start of the experiment without any axial load; this displacement may be caused by the hysteresis of the SMA wire after test actuation prior to the experiment. By subtracting the displacement caused by the external axial force with the displacement due to actuation of the wire, an improvement of 33.5% of tip rotation angle could be obtained by using the Post-Buckled Precompressed (PBP) mechanism in comparison of using the specimen without any externally applied load. An improvement of up to 50.5% of tip rotation angle could be achieved with higher current levels, but due to the fact that the control system described later could only provide a maximum actuation current of I = 0.6 A, the 0.55 A case seemed to be the right fit for the proposed application in the airfoil. Experiments also showed that the SMAPBP actuator can achieve blocked force values of up to 0.45 N. Axial buckling force measurement as a function of finite rotation angles. Measuring the initiating buckling force at finite rotation angles presented nontrivial challenges as the ensuing events were dynamic in nature (as opposed to static). Still, once the initiation force levels were applied and the structure settled in a quiescent shape, the axial buckling force could be compared to finite tip rotation levels as shown in Figure 3.

AIRFOIL

The airfoil intended to prove the concept of a SMAPBP actuator actuated airfoil was proposed to have a chord length of 150 mm and a width of 100 mm. As a basis for the modeling of the airfoil, the NACA profile series was considered. The decision was made to use the NACA 0012 profile because of the large amount of available research data for this airfoil. The symmetric airfoil was used because this paper focuses more on the capabilities of the actuator integrated in the airfoil then maximizing the aerodynamic performance (airfoil profile design). The first 60 mm of the profile were used as a D-spar composite leading edge with the last 90 mm being the actuated length. This decision led to a 40% leading edge and 60% actuator distribution in respect to the overall length of the airfoil.

Figure 4 shows the assembly of the airfoil without the skin. The airfoil structure basically consisted of four main components; the carbon fiber composite leading edge, the actuators, the trailing edge and the elastic latex skin. The leading edge was based on the shape of a NACA0012 [61] airfoil with a chord length of 150 mm. In order to obtain a smooth crossover between the leading edge contour and the straight skin to the trailing edge, a rounded edge was designed at the end of the leading edge. At a distance of 48 mm from the leading edge, a rounded chamfer was used to smoothen the crossover.

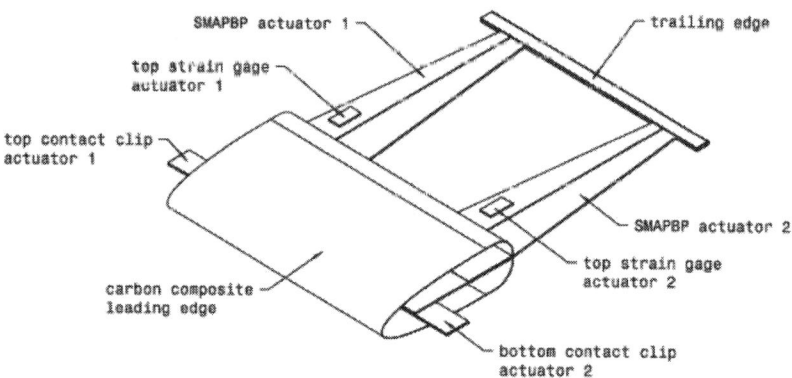

Figure 4: Drawing of assembled airfoil without applied skin.

The SMAPBP actuator was clamped and bonded to the inside of the leading edge at the ends of the carbon composite layer. This design approach made it possible to use the entire chord length of the airfoil more effectively because the clamping area of the actuator was integrated into the substructure of the leading edge and not behind the leading edge. This approach led to a higher actuated length and therefore larger trailing edge displacements.

The trailing edge consisted of a simple bended steel sheet. The trailing edge had the only purpose to connect the two actuators and to obtain a fixture for the applied skin. The trailing edge had an overall length of 100 mm (similar to the width of the leading edge), a thickness of 0.127 mm and a width of 5 mm. Figure 5 shows the manufactured airfoil with attached elastic skin.

Figure 5: Manufactured airfoil with elastic skin.

The skin had the purpose to give the airfoil a smooth surface and to apply the necessary buckling load to the SMAPBP actuator. The skin consisted of a 100 mm long and 40 mm wide latex tube. Based on tensile tests, an applied axial load of F_{askin} of the skin between 6 N and 8 N was measured, which was just below the Euler buckling loads both computed and experimentally observed.

Adaptive Shape Change of Airfoil

With simple trigonometric relationships the form change of the airfoil can be modeled mathematically. Figure 6 and Figure 7 show the entire airfoil with leading edge, actuator and skin for variable actuator tip rotation angles. The skin covers the whole outline of the airfoil. The following figures show the shape of the airfoil for actuator tip rotation angles of δ = 0° and δ = 30°.

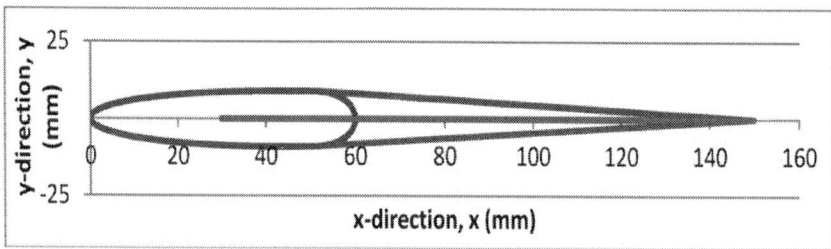

Figure 6: Airfoil cross section at δ = 0°.

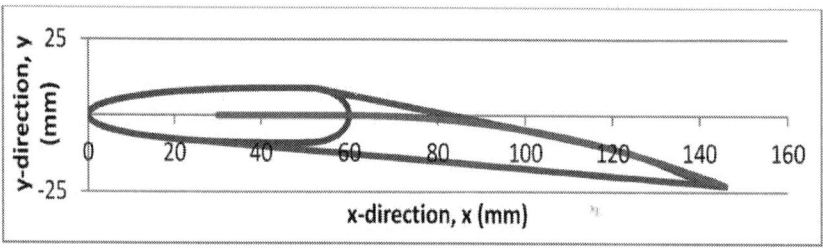

Figure 7: Airfoil cross section at δ = 30°.

Active Position Feedback System

Due to the highly hysteretic and nonlinear behavior of the shape memory alloy actuator, it was necessary to introduce an active position feedback system to the airfoil. The control system had the function to control the tip displacement over the applied actuator current. Different concepts for the feedback system were considered and the decision was made to use strain gages to measure the variation of strain on the surface of the actuator.

The strain on the surface correlates directly to the curvature of the actuator and therefore to the tip rotation angle. Figure 8 shows the principle architecture of the active position feedback system proposed for the airfoil. For the proposed airfoil the strain gage in a Wheatstone bridge is used as sensor while the actuator is the actual SMAPBP actuator with current amplifier and the controller which is the control routine is developed in Labview®.

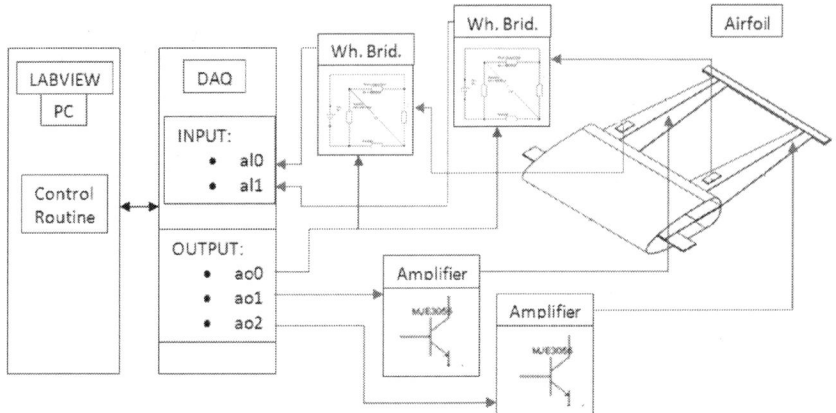

Figure 8: Position feedback system architecture.

By displacing the trailing edge, the resistance of the strain gages applied on the actuators changed. The sensor (strain gage and Wheatstone bridge) senses a change in voltage that was then read into the analog input channels of the data acquisition system (DAQ). The DAQ system was connected over a USB port to a personal computer with Labview® application. The control routine simply minimizes the difference between current and desired position.

Due to the fact that Labview® only provides a variable voltage output and that a variable current level was needed to actuate the actuator's SMA wires, a current amplifier needed to be interposed. With this simple position feedback loop it was possible to overcome the nonlinear hysteretic behavior of the SMA wire and made it possible to command the trailing edge of the airfoil to a specific predetermined displacement.

Wind Tunnel Tests

Wind tunnel tests were undertaken in the open circuit 31 × 46 cm wind tunnel at The University of Kansas, which possesses a turbulence factor of 1.1. The purpose of the wind tunnel tests was to prove the performance of the airfoil in a real life application. The wind tunnel tests should prove if the SMAPBP actuator could be able to displace the trailing edge and increase the lift due to actuation. The performance of the airfoil was evaluated for different angles of attack and different actuator tip angle rotations. Figure 9 shows the airfoil in the wind tunnel in an unactuated state before the commencement of the tests.

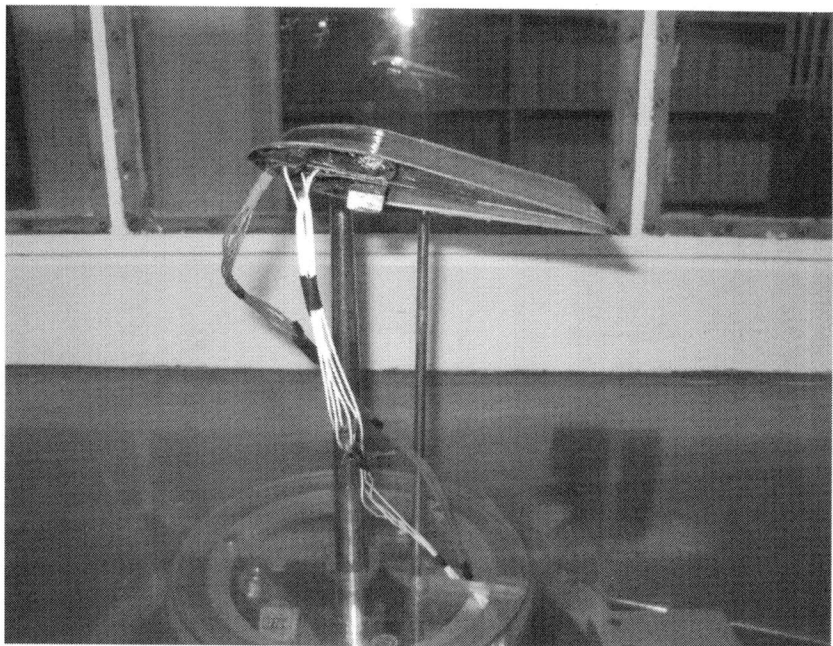

Figure 9: Side view of airfoil in wind tunnel.

Measurements were taken for actuator tip rotation angles of $\delta = 0°$, $\delta = 10°$, $\delta = 20°$ and $\delta = 30°$ with angles of attack of $\alpha = 0°$, $\alpha = 2°$, $\alpha = 6°$ and $\alpha = 10°$. For each angle of attack the measurements of the four tip rotation were undertaken. After each angle-of-attack measurement, the wind tunnel was turned off and a new angle of attack was adjusted.

Figure 10 shows the actuated airfoil; the white wires are for the actuation of the SMAPBP actuator and the brown wires are from the strain gages applied to the actuators. The SMAPBP actuator performed well during the wind tunnel tests, the trailing edge was displaced up to $\delta = 30°$. The behavior of the airfoil in the wind tunnel seemed different than without any applied air flow. A noticeable difference was seen in the power consumption of the actuators. The consumption with airflow was significantly higher than without any airflow. Without any airflow an actuator tip rotation angle of $\delta = 30°$ could be achieved with 0.6 A actuation current, during the wind tunnel tests a actuation current of up to 1.0 A needed to be applied to achieve similar displacements. This effect might be caused by convectional cooling of the SMA actuator wires by the surrounding airflow. Due to the fact that the airfoil did not have side panels, air flow could enter into the airfoil and cool down the SMA wires.

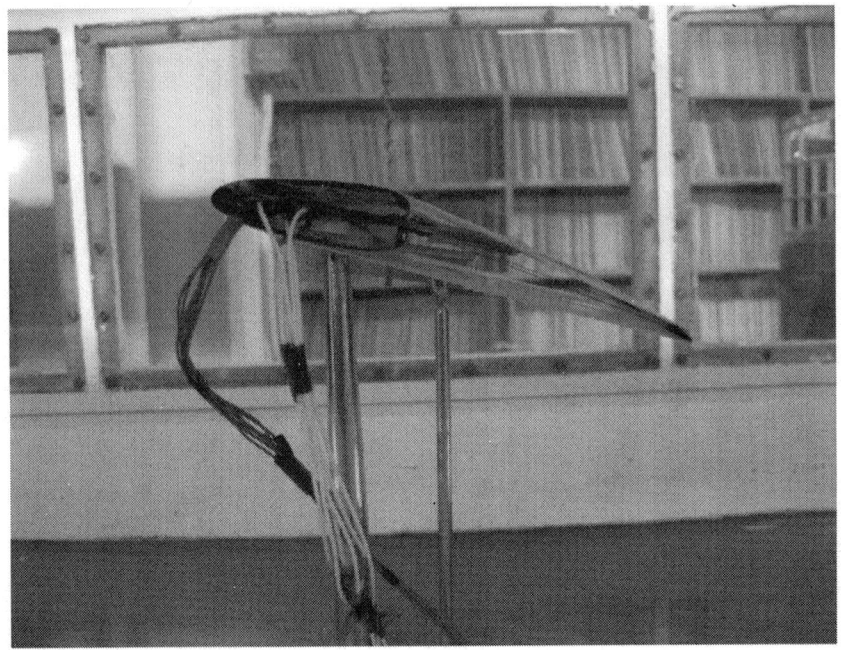

Figure 10: Side view of Airfoil in Wind Tunnel with a Tip Rotation of $\delta = 30°$.

The wind tunnel experiments obtained also lift and drag coefficients for the different actuator tip angle rotations and angles of attack.

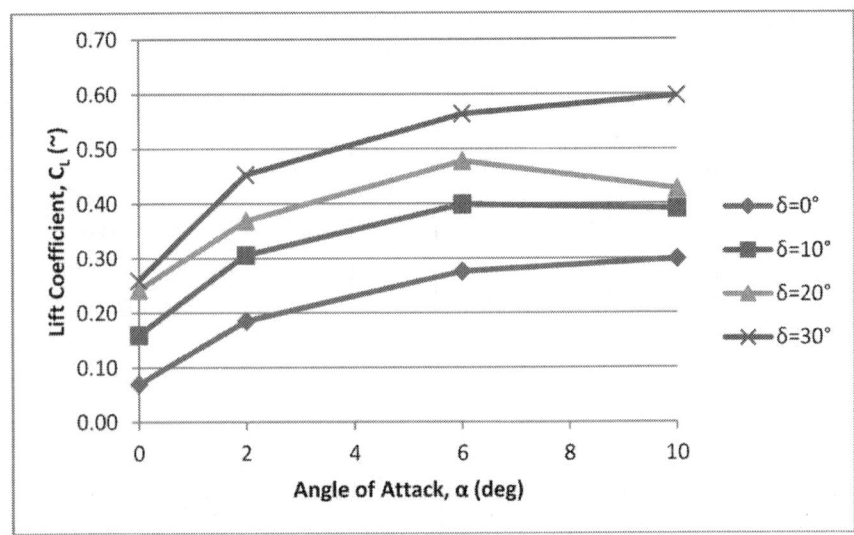

Figure 11: Lift coefficient over angle of attack from wind tunnel experiments.

Figure 11 shows the lift coefficient over various angles of attack and actuator tip displacement angles. The figure shows that the lift increased with an increasing angle of attack α and increasing actuator tip rotation angle δ. The values for the lift coefficient for an angle of attack of $\alpha = 6°$ and $\alpha = 10°$ had similar values. For an actuator tip rotation angle of $\delta = 20°$, the lift coefficient of the $\alpha = 10°$ case was even smaller than the $\alpha = 6°$.

Nevertheless, the wind tunnel tests proved that the airfoil was working under operating conditions. The experiments showed that the lift can be increased substantially by actuating the SMAPBP actuator. The wind tunnel experiments also showed that the SMAPBP actuator can overcome the aerodynamic loads and displace the trailing edge.

CONCLUSIONS

The Post-Buckled Precompressed (PBP) amplification techniques works well with antagonistically arranged SMA bender beams of the SMAPBP actuator. A 135 mm long steel-fiberglass-epoxy composite actuator was actuated by a pair of 0.254 mm diameter Dynalloy Flexinol wires. The SMAPBP actuator design presented in this paper is capable of achieving high tip displacements and rotations. The specimen demonstrated deflection levels in excess of 120° peak-to-peak, representing a deflection growth of 50.5% at axial loads of 5 N. Currently, the actuator is optimized for a buckling force of around 3.5 N-4.5 N and an actuated/buckling length of 90 mm. Therefore, the work presented in this paper can serve as a basic design study for future SMAPBP actuators for other load cases and applications. With the employment of the SMAPBP actuator in an airfoil, the concept of an adaptive flap system with SMA actuation and applied PBP mechanism was presented. Experiments showed that the SMAPBP actuator can achieve high trailing edge displacements and overcome blocked forces originating from airloads. Wind tunnel experiments showed that the airfoil was working well in flight conditions and that the SMAPBP actuator was able to overcome the aerodynamic loads introduced by the air flow.

ACKNOWLEDGMENTS

The authors would like to acknowledge the Transportation Research Institute (TRI) of the University of Kansas for its financial support of the research in the Adaptive Aerostructures Laboratory of the University of Kansas.

AUTHOR CONTRIBUTIONS

The authors contributed equally to this paper.

REFERENCES

1. Crawley, E.; Lazarus, K.; Warkentin, D. Embedded actuation and processing in intelligent materials. In Proceedings of the 2nd International Workshop on Composite Materials and Structures for Rotorcraft, Troy, NY, USA, 14–15 September 1989.
2. Lazarus, K.; Crawley, E. Multivariable active lifting surface using strain actuation: analytical and experimental results. In Proceedings of the Third International Conference on Adaptive Structures, San Diego, CA, USA, 9–11 November 1992.
3. Lazarus, K.; Crawley, E.; Bohlmann, J.D. Static aerolastic control using strain actuated adaptive structures. *J. Intell. Mater. Syst. Struct.* **1991**, *2*, 386–410.
4. Spangler, R.L.; Hall, S.R. Piezoelectric actuators for helicopter rotor control. In Proceedings of the 31st Structures, Structural Dynamics and Materials Conference, Long Beach, CA, USA, 2–4 April 1990.
5. Barrett, R. Intelligent rotor blade actuation through directionally attached piezoelectric crystals. In Proceedings of the 46th AHS National Conference and Forum, Washington, DC, USA, 21–23 May 1990.
6. Barrett, R. Intelligent Rotor Blade and Structures Development using Piezoelectric Crystals. Master's Thesis, the UM, College Park, MD, USA, 1990.
7. Barrett, R. Method and Apparatus for Sensing and Actuating in a Desired Direction. *U.S. Patent* 5,440,193, August 1995.
8. Barrett, R. Actuation strain decoupling through enhanced directional attachment in plates and aerodynamic surfaces. In Proceedings of the First European Conference on Smart Structures and Materials, Glasgow, UK, 12–14 May 1992; pp. 383–386.
9. Ehlers, S.M.; Weisshaar, T.A. Static Aeroelastic Behavior of an Adaptive Laminated Piezoelectric Composite Wing. In Proceedings of the 31st Structures, Structural Dynamics, and Materials Conference, Long Beach, CA, USA, 2–4 April 1990; pp. 1611–1623.
10. Ehlers, S.M.; Weisshaar, T.A. Adaptive wing flexural axis control. In Proceedings of the Third International Conference on Adaptive Structures, San Diego, CA, USA, 9–11 November 1992.

11. Ehlers, S.M.; Weisshaar, T.A. Effects of material properties on static aerolastic control. In Proceedings of the 33rd Structures, Structural Dynamics and Materials, Dallas, TX, USA, 13–15 April 1992.
12. Barrett, R. Active plate and missile wing development using dap elements. *AIAA J.* **1994**,*32*, 601–609.
13. Barrett, R. Active plate missile development using EDAP elements. *J. Smart Mater. Struct.* **1992**, *1*, 214226.
14. Barrett, R. Active composite torque-plate fins for subsonic missiles. In Proceedings of the Dynamic Response of Composite Structures Conference, New Orleans, LA, USA, 20 August–1 September 1993.
15. Barrett, R. *Advanced low-Cost Smart Missile Fin Technology Evaluation*; Report to the USAF Armament Directorate, Eglin Air Force Base: Eglin, FL, USA, 1993.
16. Barrett, R.; Brozoski, F.; Gross, R.S. Design and testing of a subsonic all-moving adaptive flight control surface. *AIAA J.* **1997**, *35*, 1217–1219.
17. Barrett, R.; Brozoski, F. Missile flight control using active flexspar actuators. *J. Smart Mater. Struct.* **1996**, *5*, 121–128.
18. Barrett, R. Active aeroelastic tailoring of an adaptive flexspar stabilator. *J. Smart Mater. Struct.* **1996**, *5*, 723–730.
19. Barrett, R. *Invention and Evaluation of the Barrel Launched Adaptive Munition (BLAM)*; Final report for USAF contract no. F-49620-93-C-0063, USAF Wright Laboratory Flight Vehicles Branch: Wright-Patterson AFB, OH, USA, 1995.
20. Barrett, R.; Stutts, J. *Barrel-Launched Adaptive Munition BLAM Experimental Round Research*; Final report for USAF contract no. F-49620-93-C-0063, USAF Wright Laboratory Flight Vehicles Branch: Wright-Patterson AFB, OH, USA, 1997.
21. Winchenbach, G. *Cone Aerodynamic Test*; Aeroballistic Research Facility Ballistic Spark Range Technical Report. USAF Wright Laboratory Flight Vehicles Branch: Wright-Patterson AFB, OH, USA, 1996.
22. Barrett, R.; Stutts, J. Modelling, Design and testing of a barrell-launched adaptive munition. In Proceedings of the 4th Annual SPIE Symposium on Smart Structures and Materials, San Diego, CA, USA, 3–6 March 1997.
23. Barrett, R. *Design, Construction and Testing of a Prrof-of-Cncept Smart Compressed Reversed Adaptive Munition*; SCRAM Report, Final Report to the USAF Armament Directorate, Wright Laboratory: Eglin AFB, FL, USA, 1996.
24. Barrett, R.; Stutts, J. Development of a piezoccramic flight control surface actuator for highly compressed munitions. In Proceedings of the 39th Structures, Structural Dynamics and Materials Conference, Long Beach, CA, USA, 20–23 April 1998.

25. Barrett, R. *Construction and Test Report for a Rotationally Active Linear Actuator (RALA) Adaptive Canard*; Final Report for McDonnell Douglas: St. Louis, MO, USA, 1997.
26. Knowles, G.; Barrett, R.; Valentino, M. Self-contained high authority control of miniature flight control systems for aera dominace. In Proceedings of the SPIE 11th International Symposium on Smart Structures and Materials, San Diego, CA, USA, 14–18 March 2004.
27. Lee, G. *Range-Extended Adaptive Munition (REAM)*; Final Report from Lutronix Corporation to the DARPA: Del Mar, CA, USA, 1999.
28. Lee, G. *40/50 Caliber Range-Extended Adaptive Munition (REAM)*; Final Report from Lutronix Corporation to the US Army TACOM-ARDEC: Del Mar, CA, USA, 2000.
29. Rabinovitch, O.; Vinson, J.R. On the design of piezoelectric smart fins for flight vehicles.*J. Smart Mater. Struct.* **1993**, *12*, 686–695.
30. Barrett, R. *High Bandwith Electric Rotor Blade Actuator Study*; Phase I SBIR proposal submitted to the U.S. Army Aviation System Command: St. Louis, MO, USA, 27 June 1992.
31. Barrett, R.; Stutts, J. Design and testing of a 1/12th scale solid state adaptive rotor. *J. Smart Mater. Struct.* **1997**, *6*, 686–695.
32. Barrett, R.; Frye, P.; Schliesman, M. Design, constructionand characterization of a flightworty piezoelectric solid state adaptive rotor. *J. Smart Mater. Struct.* **1998**, *7*, 422–431.
33. Lee, G. *Design and Testing of the Kolibri Vertical Take-Off and Landing Micro Aerial Vehicle*; Final report to the Department of Defense CounterDrug Technology Office: Washington, DC, USA, 1997.
34. Barrett, R.; Howard, N. Adaptive aerostructures for subscale aircraft. In Proceedings of the 20th Southeastern Conference on Theoretical and Applied Mechanics, Pine Mountains, GA, USA, 17 April 2000.
35. Lesieutre, G.A.; Davis, C.L. Can a Coupling Coefficient of a Piezoelectric Actuator be Higher than its Active Material? *J. Intell. Mater. Syst. Struct.* **1997**, *8*, 859–867.
36. Lesieutre, G.A.; Davis, C.L. Transfer Having a Coupling Coefficient Higher than its Active Material. *U.S. Patent* 6,236,143, 22 May 2001.
37. Barrett, R.; Tisco, P. PBP Adaptive Actuator Device and Embodiments. *International Patent* Application Number PCT/NL2005/000054, via TU Delft. 18 February 2005.
38. Barrett, R.; Vos, R.; Tisco, P.; De Breuker, R. Post-Buckled Precompressed (PBP) Actuators: Enhancing VTOL Autonomous High Speed MAVs. In Proceedings of the 13th AIAA/ASME/AHS Adaptive Structures Conference, Austin, TX, USA, 18–21 April 2005.

39. Vos, R.; De Breuker, R.; Barrett, R.; Tisco, P. Morphing Wing Flight Control via Post-Buckled Precompressed Piezoelectric Actuators. *AIAA J. Aircr.* **2007**, *44*, 1060–1068.

40. Barrett, R.; McMurtry, R.; Vos, R.; Tisco, P.; De Breuker, R. Post-Buckled Precompressed Pieyoelectric Flight Control Actuator Design, Development and Demonstration. *J. Smart Mater. Struct.* **2006**, *15*, 1323–1331.

41. De Breuker, R.; Vos, R.; Barrett, R.; Tisco, P. Nonlinear Semi-Analytical Modeling of Post-Buckled Precompressed (PBP) Piezoelectric Actuators for UAV Flight Control. In Proceedings of the 47th AIAA/ASME/ASCE/AHS/ASC Structures, Structural Dynamics and Materials Conference 14th AIAA/ASME/AHS Adaptive Structures Conference, Newport, RI, USA, 1–4 May 2006.

42. Van Schravendijk, M.; Groen, M.; Vos, R.; Barrett, R. Closed Loop Control for High Bandwidth, High Curcature Post-Buckled Precompressed Actuators. In Proceedings of the 50th AIAA/ASME/ASCE/AHS/ASC Structures, Structural Dynamics, and Materials Conference 17th AIAA/ASME/AHS Adaptive Structures Conference 11th AIAA No, Palm Springs, CA, USA, 4–7 May 2009.

43. Arrieta, A.F.; Bilgen, O.; Friswell, M.I.; Hagedorn, P. Passive load alleviation bi-stable morphing concept. *AIP Adv.* **2012**, *2*.

44. Bilgen, O.; Friswell, M.I. Implementation of a Continuous-Inextensible-Surface Piezocomposite Airfoil. *J. Aircr.* **2013**, *50*, 508–518.

45. Friswell, M.I.; Ali, S.F.; Bilgen, O.; Adhikari, S.; Lees, A.W.; Litak, G. Nonlinear Piezoelectric Vibration Energy Harvesting from an Inverted Cantilever Beam with Tip Mass. *J. Intell. Mater. Syst. Struct.* **2012**, *23*, 1505–1521.

46. Arrieta, A.F.; Bilgen, O.; Friswell, M.I.; Hagedorn, P. Morphing dynamic control for bi-stable composites. *J. Intell. Mater. Syst. Struct.* **2012**.

47. Bilgen, O.; Butt, L.M.; Day, S.R.; Sossi, C.A.; Weaver, J.P.; Wolek, A.; Mason, W.H.; Inman, D.J. A Novel Unmanned Aircraft with Solid-State Control Surfaces: Analysis and Flight Demonstration. *J. Intell. Mater. Syst. Struct.* **2012**.

48. Borowiec, M.; Litak, G.; Friswell, M.I.; Ali, S.F.; Adhikari, S.; Lees, A.W.; Bilgen, O. Energy Harvesting in Piezoelastic Systems Driven by Random Excitations. *Int. J. Struct. Stab. Dyn.* **2012**.

49. Arrieta, A.F.; Bilgen, O.; Friswell, M.I.; Hagedorn, P. Modeling and configuration control of multi-stable piezoelectric composites. *Smart Mater. Struct.* **2013**. submitted for publication.

50. Bilgen, O.; Wang, Y.; Inman, D.J. Electromechanical Comparison of Cantilevered Beams with Multifunctional Piezoceramic Devices. *Mechan. Syst. Signal Process.* **2012**, *27*, 763–777.

51. Karami, M.A.; Bilgen, O.; Inman, D.J.; Friswell, M.I. Experimental and Analytical Parametric Study of Single Crystal Unimorph Beams for Vibration Energy Harvesting.*Trans. Ultrason. Ferroelectr. Freq. Control* **2011**, *58*, 1508–1520.

52. Bilgen, O.; Inman, D.J.; Friswell, M.I. Theoretical and Experimental Analysis of Hysteresis in Piezocomposite Airfoils Using the Classical Preisach Model. *J. Aircr.* **2011**,*48*, 1935–1947.

53. Barbarino, S.; Bilgen, O.; Ajaj, R.M.; Friswell, M.I.; Inman, D.J. A Review of Morphing Aircraft. *J. Intell. Mater. Syst. Struct.* **2011**, *22*, 823–877.

54. Bilgen, O.; Karami, M.A.; Inman, D.J.; Friswell, M.I. The Actuation Characterization of Cantilevered Unimorph Beams with Single Crystal Piezoelectric Materials. *Smart Mater. Struct.* **2011**, *20*.

55. Bilgen, O.; De Marqui Junior, C.; Kochersberger, K.B.; Inman, D.J. Macro-Fiber Composite Actuators for Flow Control of a Variable Camber Airfoil. *J. Intell. Mater. Syst. Struct.* **2011**, *22*, 81–91.

56. Bilgen, O.; Kochersberger, K.B.; Inman, D.J.; Ohanian, O.J. Macro-Fiber Composite Actuated Simply-Supported Thin Morphing Airfoils. *Smart Mater. Struct.* **2010**, *19*.

57. Bilgen, O.; Erturk, A.; Inman, D.J. Analytical and Experimental Characterization of Macro-Fiber Composite Actuated Thin Clamped-Free Unimorph Benders. *J. Vib. Acoust.***2010**, *132*.

58. Bilgen, O.; Kochersberger, K.B.; Inman, D.J.; Ohanian, O.J. Novel, Bi-Directional, Variable Camber Airfoil via Macro-Fiber Composite Actuators. *J. Aircr.* **2010**, *47*, 303–314

59. Bilgen, O.; De Marqui Junior, C.; Kochersberger, K.B.; Inman, D.J. Piezoceramic Composite Actuators for Flow Control in Low Reynolds Number Airflow. *J. Intell. Mater. Syst. Struct.* **2010**, *21*, 1201–1212.

60. Dynalloy, I. Flexinol™ Technical Data. Available online: http://www.robotshop.com/media/files/pdf/flexinol-technical-data.pdf (accessed on 27 July 2015).

61. Ladson, C.L.; Brooks, C.W., Jr.; Hill, A.S.; Sproles, D.W. Computer program to obtain ordinates for NACA airfoils. *NASA Technical Memorandum*; Available online: http://ntrs.nasa.gov/archive/nasa/casi.ntrs.nasa.gov/19970008124.pdf (accessed on 27 July 2015).

CITATION

Thomas Sinn, and Ron Barrett, Design, Manufacturing and Test of a High Lift Secondary Flight Control Surface with Shape Memory Alloy Post-Buckled Precompressed Actuators, doi:10.3390/act4030156

CHAPTER 6

Numerical Study on Flow Fields and Aerodynamics of Tilt Rotor Aircraft in Conversion Mode Based on Embedded Grid and Actuator Model

Zhang Ying, Ye Liang, Yang Shuo

Aviation Key Laboratory of Science and Technology on Aerodynamics of High Speed and High Reynolds Number, AVIC Aerodynamics Research Institute, Shenyang 110034, China

ABSTRACT

A method combining rotor actuator disk model and embedded grid technique is presented in this paper, aimed at predicting the flow fields and aerodynamic characteristics of tilt rotor aircraft in conversion mode more efficiently and effectively. In this method, rotor's influence is considered in terms of the momentum it impacts to the fluid around it; transformation matrixes among different coordinate systems are deduced to extend actuator method's utility to conversion mode flow fields' calculation. Meanwhile, an embedded grid system is designed, in which grids generated around fuselage and actuator disk are regarded as background grid and minor grid respectively, and a new method is presented for 'donor searching' and 'hole cutting' during grid assembling. Based on the above methods, flow fields of tilt rotor aircraft in conversion mode are simulated, with three-dimensional Navier–Stokes equations discretized by a second-order upwind finite-volume scheme and an implicit lower–upper symmetric Gauss–Seidel (LU-SGS) time-stepping scheme. Numerical results demonstrate that the proposed CFD method is very effective in simulating the conversion mode flow fields of tilt rotor aircraft.

INTRODUCTION

Tilt rotor aircraft is a flight vehicle which can land and take off vertically, as well as convert to propeller mode to achieve high-speed cruise performance. Extensive researches, covering a number of areas including aerodynamic interactions, performance, aeroacoustics and dynamics, have been conducted on it in recent years due to its unique advantages and technical complexity. In the aspect of experimental research, various experiments have been carried out by NASA Ames Research Center, Bell Helicopters Inc.[1, 2 and 3] and other institutions. Abundant experimental data has been obtained, which provides great assistance to the tilt rotor aircraft design and manufacture. In the aspect of computational research, due to the complexities of tilt rotor flow fields and its aerodynamic characteristics, more and more researchers started to study tilt rotor aircraft aerodynamics by solving Navier–Stokes equations instead of using conventional methods. For example, in the mid-1990s, the flow field of V-22 rotor/wing configuration in hover was simulated by Meakin[4] with the unsteady thin-layer Navier–Stokes equations, and rotor motion and rotor/airframe interference were simulated directly using moving body overset grid methods. The flow over a wing-fuselage-nacelle configuration of the V-22 tilt rotor aircraft in forward flight was simulated by Tai[5] with a multi-zone, thin-layer Navier–Stokes method, and major flow features including the three-dimensional flow separation due to viscous-vortex interactions observed experimentally were captured. After the year 2000, overset-grid Reynolds Averaged Navier–Stokes (RANS) solvers, OVERFLOW and Fun3D, were used by Gupta and Baeder[6] and Lee-Rausch[7] respectively to numerically simulate the flow fields of a simplified Quad Tilt Rotor (QTR) in forward flight and an isolated tilt rotor in hover. Aerodynamic characteristics of tilt rotor in hover were computed by Potsdam and Strawn[8] by solving the governing equations in rotation reference frame. More detailed investigations on tilt rotor aircraft were presented in the doctorate thesis of Gupta.[9] In his paper, flow fields around a simplified Quad Tilt rotor, both in and out of ground effect, were simulated; rotor was modeled by both actuator disk and body conforming meshes method, the results obtained by these two methods were then compared.

A feature of tilt rotor aircraft is the existence of conversion mode. In this status, the manipulation and dynamic response of its blades are more complex than those of conventional helicopter rotors or propeller because of the more severe wake distortion and unsteady blade loads. Research on tilt rotor aircraft in conversion mode is still a challenge until now, although extensive research works on this aircraft in helicopter or propeller mode have been conducted. Challenges to the research work on tilt rotor aircraft arise from the complexity of flow physics in conversion mode, which is reflected in the following two aspects. One is that the rotor wake is intensely distorted due to the blade manipulation for aerodynamic balance and the other is that flow fields induced by wing and rotors are closely coupled and severe aerodynamic interference phenomenon appears, with the reason that the forward flight speed is usually not very high and the space between rotor blades and wings is quite narrow.

Considering the importance and complexity of the research on conversion mode, this paper mainly focuses on numerical methods to be used for conversion mode simulation. In general, accurate modeling of flow around individual rotor blades is a complex task even for a conventional helicopter configuration, which means much more time and resource requirements for tilt rotor aircraft conversion mode calculation. As the first step of the research plan, this paper proposes a new approach by combining rotor actuator disk model with embedded grid technique, aimed at providing a rapid, effective and universal method, which can be conveniently transplanted to the rotor blade body-fitted flow field calculation in future. The work in this paper can be described as follows: rotor is modeled as an actuator disk in order to ameliorate computational time costs and make simulation works possible, and transformation matrixes among different coordinate systems are deduced so that the actuator model can be used at different tilt angles. Meanwhile, embedded grid system is designed to describe relative movement among the rotors and wing/fuselage, and a new method is presented for 'donor searching' and 'hole cutting' during grid assembling. Grids generated around fuselage and actuator disk are regarded as background grid and minor grid respectively, and the influence of actuator disk on flow fields at different tilt angles is simulated by the location change of minor gird.

A CFD solver is put forward after comprehensively analyzing the requirements of flow field calculation for tilt rotor configurations in conversion mode, by combining methods mentioned above and Navier–Stokes equations solving technique. Three-dimensional Navier–Stokes equations are discretized using a second-order upwind finite-volume scheme and an implicit LU-SGS scheme. Induced velocity fields of experimental rotors are simulated firstly for validation purpose, and then flow fields of a tilt rotor aircraft model are calculated. Results indicate that the present CFD method is very effective and has the ability to capture essential flow field characteristics and to predict aerodynamic interactions of tilt rotor aircraft in conversion mode.

METHODOLOGY

Grid System

An embedded grid system with the rotor modeled as actuator disk is adopted in the present work. Considering the relative movement among different components of tilt rotor configurations, a minor grid with the rotor modeled as actuator disk and a background grid enclosing unmoving bodies (fuselage) are generated separately, and the minor grid is fully embedded inside the background grid. Taking the universality into consideration, topology of grid cells is not restricted (although the pure hexahedron cells are used in the following illustrations).

Preprocessing of overset grid conventionally involves two steps: 'hole cutting' and 'donor searching'. The related previous investigations are extensive and various methods are developed, such as explicit and implicit methods for hole cutting,[10, 11 and 12] inverse map,[13] neighbor-to-neighbor (N2N)[14] and alternating digital tree (ADT) searching algorithm[15] for donor cell searching. Based on these previous studies, a more adaptable scheme is designed and applied in the present work to meet the requirements for tilt rotor aircraft calculation as described in detail as follows.

(1) For each grid block, cells which are directly adjacent to body surface or actuator disk are marked as layer 1; those unmarked cells which are adjacent to layer 1 are then marked as layer 2. Then, the left cells will be marked with different layer numbers according to the method, and these layer numbers will be used as adaptable parameters for adjusting range of 'hole cutting'.

(2) For each cell (noted as C) in one grid block, cell in other grid blocks which contains the cell center of C is searched and noted as donor cell (D). If the layer number of cell D is less than or equal to a given constant, cell C will be regarded as 'Hole Cell', which should be blanked out during computation. Those cells which do not have donor cells are temporarily considered as normal cells. Neighbor-to-Neighbor (N2N) searching algorithm is adopted for donor cell searching and the efficiency of this algorithm relies on the start location of donor cell. For problems concerning grid motion, recalculation of the domain connectivity, including hole cutting and interpolation coefficients at each time step is required, and information from the previous time step is reused instead of domain connectivity solutions from scratch to enable speed-ups. Furthermore, as we know neighbor-to-neighbor algorithm performs well only when the domain is continuous and convex, so in order to satisfy this condition,

auxiliary gird is generated inside body and the convexity of outer boundary of each grid block must be ensured.

(3) For loop over all normal cells, the cell whose neighbors have 'Hole Cell' is defined as 'Interpolated Cell'.

A multi-element airfoil model (as shown in Fig. 1) is used to better elucidate this embedded grid technique. Gird enclosing the main wing is defined as background grid, while grids around leading-edge slat and flap are defined as minor grids. It can be seen from this figure that some minor grids cells (some of them are outer boundary cells) overlapping in-body region of background grid must be blanked out. Remaining outer boundary of the minor grid blocks gets information from background grid, which acts as a boundary condition for the minor grid. Meanwhile, holes are cut in the background grid in those regions where the finer near body grid (minor grid) is available; since hole is cut at the background grid, an inner boundary is generated in the background grid and those cells need to receive information from the minor grid. A typical searching path adopted to find donor cell for an arbitrary cell is shown in Fig. 2 and the arrows demonstrate the searching direction. Fig. 3 is the grids after 'hole cutting' and Fig. 4 shows background grid after hole cutting by using different layer numbers.

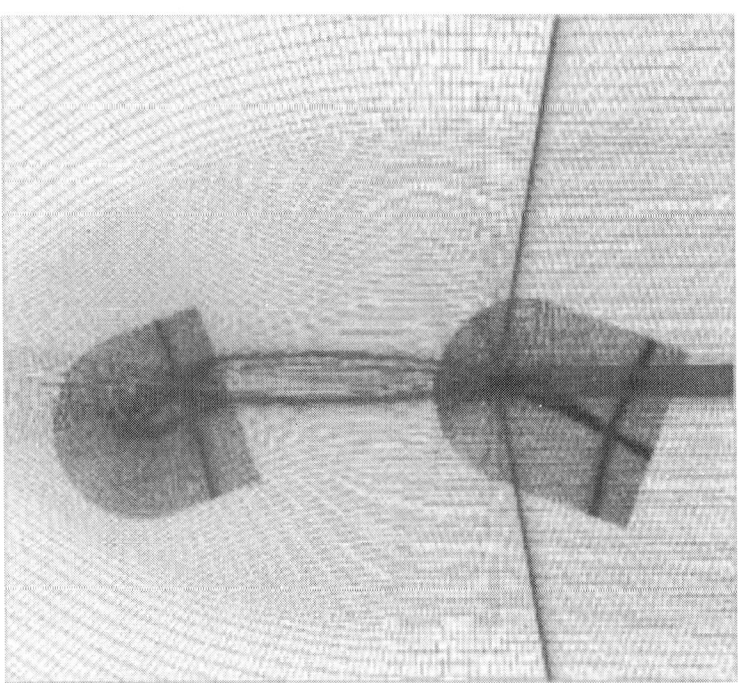

Figure 1: Grids before hole cutting.

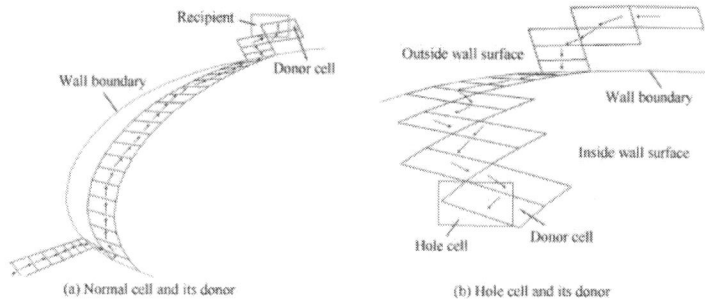

(a) Normal cell and its donor   (b) Hole cell and its donor

Figure 2: Searching path adopted to find the donor cell for an arbitrary cell.

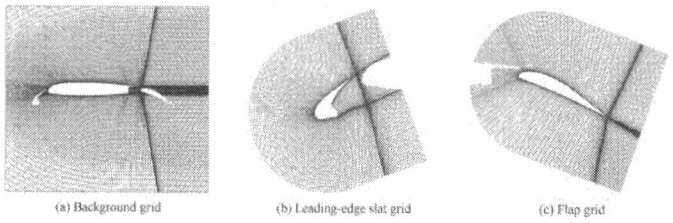

(a) Background grid   (b) Leading-edge slat grid   (c) Flap grid

Figure 3: Grids after hole cutting.

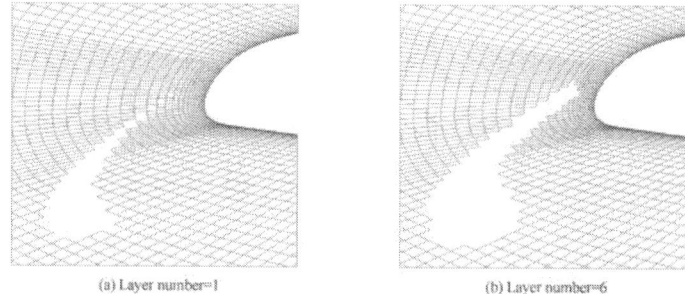

(a) Layer number=1   (b) Layer number=6

Figure 4: Background grid after hole cutting with different layer numbers' selection.

NUMERICAL METHOD

Governing Equations

For a three-dimensional flow through a finite volume Ω moving with a speed V_t, which is enclosed by the boundary surface $\partial\Omega$ and an exterior normal vector n, integral form of the conservation equations in inertial reference frame is given as

$$\frac{\partial}{\partial t} \int \int \int W d\Omega + \oiint_{\partial\Omega}(F - WV_t) \cdot n dS = Q \tag{1}$$

where W denotes the vector of conservative variables, F convective flux vector, and Qsource term. These vectors are given below:

$$\begin{cases} W = [\rho, \rho V, \rho E]^T \\ F = [\rho V \cdot n, \rho V(V \cdot n) \cdot n + Pn, \\ \rho E(V \cdot n) + PV \cdot n]^T \end{cases}$$

where ρ denotes the density, V vector of velocity, E total energy per unit mass and Ppressure.

Definition and Transformation of Coordinate Systems

Three different coordinate systems are adopted for describing the movement of a tilt rotor aircraft: inertial reference frame, nacelle reference frame and rotor reference frame (seeFig. 5). The inertial reference frame convention used in the present work is given in detail as follows: z-axis: longitudinal fuselage which is positive in downstream direction; y-axis: vertical direction of fuselage which is positive upward; x-axis: spanwise axis (parallel to the tilt axis of nacelle).

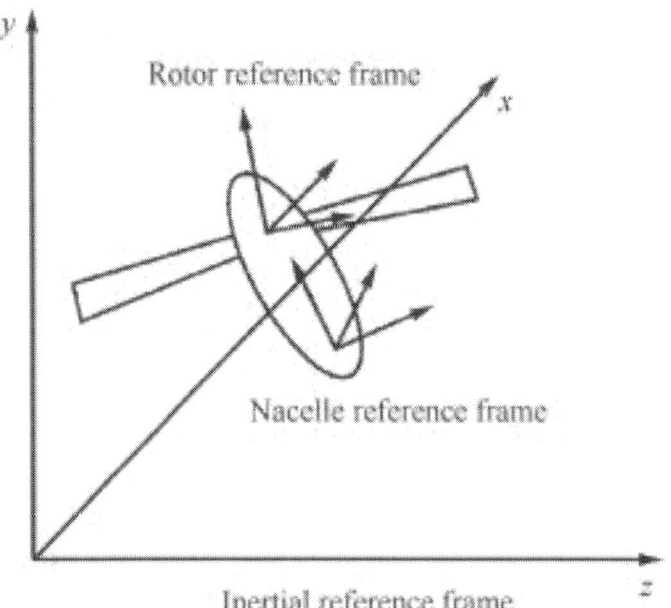

Figure 5: Reference coordinate systems definition.

Transformation matrixes between different coordinate systems are given as follows:

$$
M_1 = \begin{bmatrix} 1 & 0 & 0 \\ 0 & \cos\gamma & \sin\gamma \\ 0 & -\sin\gamma & \cos\gamma \end{bmatrix} \quad M_2 = \begin{bmatrix} \cos\psi & 0 & -\sin\psi \\ 0 & 1 & 0 \\ \sin\psi & 0 & \cos\psi \end{bmatrix}
$$

$$
M_3 = \begin{bmatrix} 1 & 0 & 0 \\ 0 & \cos\beta & \sin\beta \\ 0 & -\sin\beta & \cos\beta \end{bmatrix} \quad M_4 = \begin{bmatrix} \cos\theta & \sin\theta & 0 \\ -\sin\theta & \cos\theta & 0 \\ 0 & 0 & 1 \end{bmatrix} \tag{2}
$$

where γ, ψ, β and θ are tilt angle, azimuthal angle, flap angle and pitch angle respectively.

Coordinate transformation between inertial reference frame and rotor reference frame can be written as

$$
\begin{cases}
\begin{bmatrix} x - x_{nrc} \\ y - y_{nrc} \\ z - z_{nrc} \end{bmatrix} = [M_1][M_2][M_3][M_4] \begin{bmatrix} \xi - \xi_{rrc} \\ \eta - \eta_{rrc} \\ \zeta - \zeta_{rrc} \end{bmatrix} \\
\qquad\qquad\qquad + [M_1] \begin{bmatrix} \xi'_{rrc} - \xi'_{nrc} \\ \eta'_{rrc} - \eta'_{nrc} \\ \zeta'_{rrc} - \zeta'_{nrc} \end{bmatrix} \\
\begin{bmatrix} \xi - \xi_{rrc} \\ \eta - \eta_{rrc} \\ \zeta - \zeta_{rrc} \end{bmatrix} = [M_4]^T[M_3]^T[M_2]^T[M_1]^T \begin{bmatrix} x - x_{nrc} \\ y - y_{nrc} \\ z - z_{nrc} \end{bmatrix} \\
\qquad\qquad\qquad - [M_4]^T[M_3]^T[M_2]^T \begin{bmatrix} \xi'_{rrc} - \xi'_{nrc} \\ \eta'_{rrc} - \eta'_{nrc} \\ \zeta'_{rrc} - \zeta'_{nrc} \end{bmatrix}
\end{cases} \tag{3}
$$

where (x, y, z), (ξ, η, ζ) and (ξ', η', ζ') are coordinates in inertial reference frame, rotor rotating reference frame and nacelle rotating reference frame respectively; the subscript rrc and nrc denote rotor rotation center and nacelle rotation center.

Rotor Modeling Method

The rotor is modeled as an infinite thin disk[16] and there are two aspects which need to be considered in determining the rotor's influence on the flow field. One aspect is to find the locations in the physical or computational domain where the influence is felt. The other aspect is to determine the momentum equation source terms Q in Eq. (1) at these locations.

The actuator disk is treated as a separate plane in the flow field and discretized into small elements. For each element, there are two grid cells, locating above and below the element respectively. The force acting on an arbitrary element is denoted as $(S_x, \quad S_y, \quad S_z)$ and distributed equally into the two grid cells. Consequently, the source term (MS) added to the governing equations that is induced by the rotor's influence is written as

$$(MS)_K = \left[0, \quad S_x/2, \quad S_y/2, \quad S_z/2, \quad (uS_x + vS_y + wS_z)/2 \right]^T$$

where u, v, w denote velocity components.

Take a tiny element on the plane of actuator disk as the investigation object, which is located at a distance r from the center of hub, with a length of dr along the spanwise direction and a width of $\Delta\theta$ along the circumferential direction. Supposing that the force acting on the blade is dF the instantaneous force acting on the fluid element at this location is then $-dF$ and the resultant force is $-NdF$ for N blades. Consequently, the time averaged force acting on the surface of grid cell, whose area is S_Δ, is $[NS_\Delta(-dF)/2\pi r]dr$. The rotational speed of rotor is ω and flow velocity experienced by the airfoil in rotor reference frame is shown in Fig. 6 (Rotor rotation direction is clockwise or counter-clockwise when viewed from above the model), where α_1 is the effective angle of attack and θ_s is the initial collective angle.

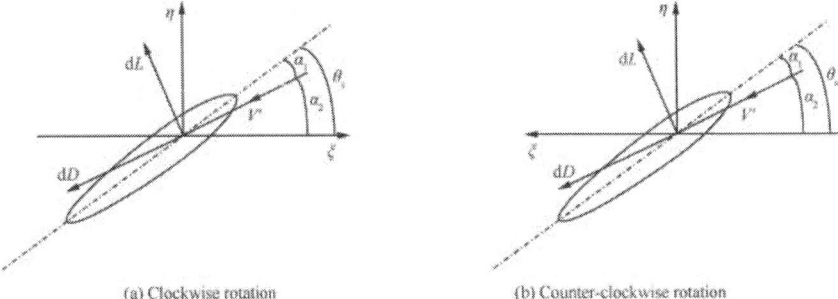

(a) Clockwise rotation   (b) Counter-clockwise rotation

Figure 6: Flow velocity experienced by the airfoil.

Since there are no aerodynamic forces along the span (ζ direction), only forces along ζ and η directions need to be calculated (see Fig. 6). For a clockwise rotating rotor, the forces $f_{\Delta\xi}$ and $f_{\Delta\eta}$ acting on the surface of grid cell, along ξ and η directions, are written in Eq. (5).

$$
\begin{cases}
f_{\Delta\eta} = \dfrac{N(-dF_\eta)S_\Lambda}{2\pi dr} = \dfrac{-N(dL\cos\alpha_2 + dD\sin\alpha_2)S_\Lambda}{2\pi r dr} \\[2mm]
\quad = \dfrac{-N\left(\frac{1}{2}\rho V'^2 C_L c dr\cos\alpha_2 + \frac{1}{2}\rho V'^2 C_D c dr\sin\alpha_2\right)S_\Lambda}{2\pi r dr} \\[2mm]
\quad = \dfrac{-N\rho V'^2 c S_\Lambda(C_L\cos\alpha_2 + C_D\sin\alpha_2)}{4\pi r} \\[3mm]
f_{\Delta\xi} = \dfrac{N(-dF_\xi)S_\Lambda}{2\pi dr} = \dfrac{-N(-dL\sin\alpha_2 + dD\cos\alpha_2)S_\Lambda}{2\pi r dr} \\[2mm]
\quad = \dfrac{-N\left(-\frac{1}{2}\rho V'^2 C_L c dr\sin\alpha_2 + \frac{1}{2}\rho V'^2 C_D c dr\cos\alpha_2\right)S_\Lambda}{2\pi r dr} \\[2mm]
\quad = \dfrac{-N\rho V'^2 c S_\Lambda(-C_L\sin\alpha_2 + C_D\cos\alpha_2)}{4\pi r}
\end{cases}
$$

$$(5)$$

where ρ is the density changing along with the flow fields' iterating, r the radial location, c the blade chord-length, S_Λ the area of grid cell, and V' relative resultant velocity in rotor reference frame; dL and dD are lift and drag acting on the blade, and C_L and C_D are lift and drag coefficient. $\alpha_2 = \arctan(V_\eta/V_\xi)$, where V_ξ and V_η are the components of V' along ξ direction and η direction respectively. Similar equations can be deduced for counter-clockwise rotating rotor.

Method for Flux Computation
ROE's scheme[17] is used in spatial discretization and flux function on the interface can be written as

$$
F_{i+\frac{1}{2}}^{\text{ROE}} = \frac{1}{2}\left[F_i + F_{i+1} - (W_i + W_{i+1})V_t - |A - V_t I|_{i+\frac{1}{2}}^* \Delta W\right]
$$

$$(6)$$

where $|\cdot|^*$ is the Jacobian matrix of ROE's scheme; subscript i, $i + 1$ and $i + \frac{1}{2}$ represent left cell, right cell and the interface respectively. $\Delta W = W_{i+1} - W_i$, I is identity matrix.

2.2.5. Precision of numerical scheme
The numerical scheme is of second-order accuracy, which can be derived using the gradient reconstruction technique, and the gradient of flow field variable Φ at cell centers can be calculated by the Green-Gauss law:

$$\nabla \Phi_{c0} = \frac{1}{V_i} \sum_f \Phi_f A_f$$

(7)

where Φ_f is the value of Φ at the interface center, V_i the volume of the cell i, A_f the area of the interface and the subscript c0, the center of the control volume. Meanwhile, the limiter of Venkatakrishnan [18] is used here to avoid generating new extrema.

Temporal Discretization
A Matrix-Free implicit LU-SGS method[19] for temporal discretization is used, which containsForward sweep:

$$\Delta W_i^* = D^{-1} \left[R_i - \sum_{j:j<i} \frac{1}{2} (\Delta F(W_j^*, n_{ij}) - |\lambda_{ij}| \Delta W_j^*) |S_{ij}| \right]$$

(8)

Backward sweep:

$$\Delta W_i = \Delta W_i^* - D^{-1} \sum_{j:j>i} \frac{1}{2} (\Delta F(W_j, n_{ij}) - |\lambda_{ij}| \Delta W_j) |S_{ij}|$$

(9)

where W_i (or W_i^*) and W_j (or W_j^*)denote the conservative variables of the cell i and j, ΔW_i (or ΔW_i^*) and ΔW_j (or ΔW_j^*) denote the increment of conservative variables for cell i and j, ΔF represents the increment of convective flux, the diagonal matrix is $D_i = \frac{V_i}{\Delta \tau} + \frac{1}{2} \sum_j |\lambda_{ij}| |S_{ij}|$, R_i is the residual after the m_{th} iteration in pseudo-time domain,$|\lambda ij|$ is the spectral radius of the flux Jacobian matrix and $|S ij|$ represents the area of the interface. Fig. 7 is the schematic of tilt rotor flow field calculation.

RESULTS AND DISCUSSION

Induced Velocity (Dynamic Pressure) Calculations for an Experimental Rotor

The experimental rotor model has two untwisted and untapered blades ($R = 0.914$ m) and airfoil section is NACA0012. Operating condition is: blade tip Mach number$M_{tip}=0.3285$ and collective angle $\theta_0=11°$. Fig. 8 is the comparison between predicted dynamic pressure distributions and experimental data [20] at different axial positions, where r and y represent the radial position and axial position. It can be seen from the figure that the calculated results and experimental data agree well.

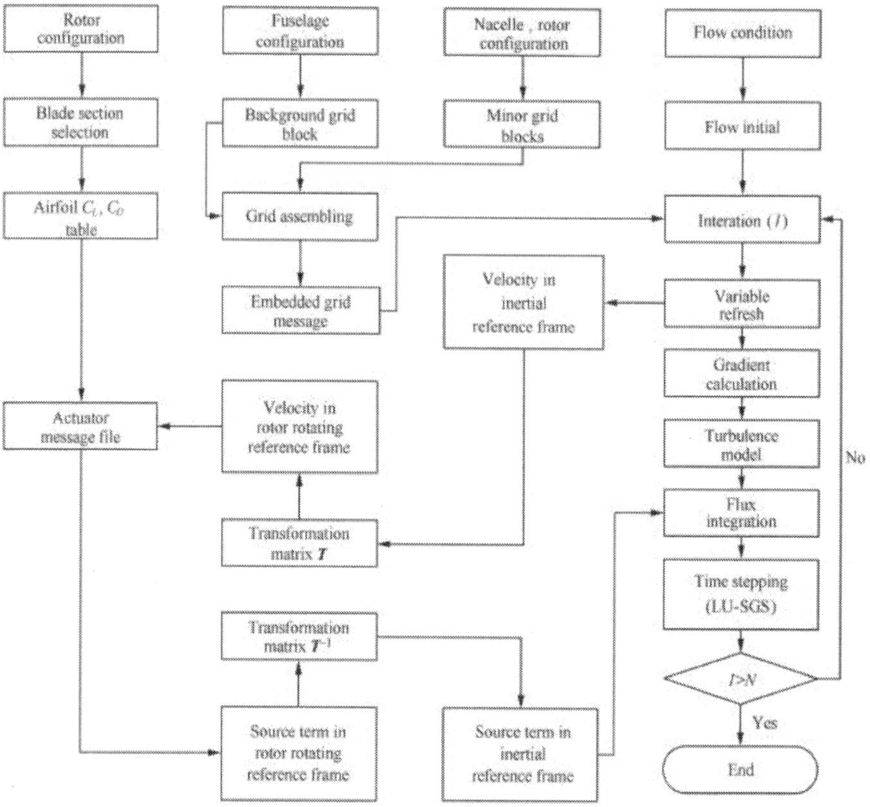

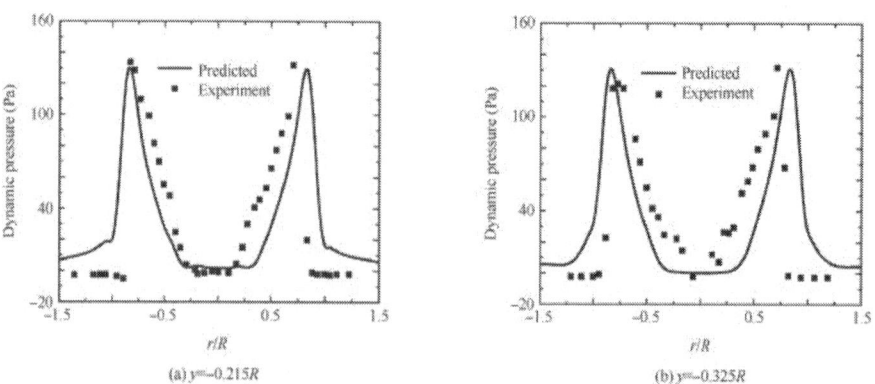

Figure 7: Schematic of tilt rotor flow field calculation.

Figure 8: Comparisons of predicted dynamic pressure distributions with experiment data.

TILT ROTOR FLOW FIELDS AND AERODYNAMIC CHARACTERISTICS

Tilt Rotor Grid System

The geometry of tilt rotor used in this paper is shown in Fig. 9, with the rotor modeled as an actuator disk. In order to conveniently describe the relative movement and guarantee grid quality, structured grids are separately generated around fuselage and nacelle which are regarded as background grid and minor grid respectively. The minor grid is totally embedded into background grid, as shown in Fig. 10. The total number of grid cells is approximately 1.5×10^7. As the geometry has a small gap between the wing and nacelle, special care was taken during grid generation to ensure enough overlap region. Meanwhile, grid refinement was made around actuator disk in order to capture dominant flow features and the maximum spacing is less than $0.01c$ (c denotes the airfoil chord length at blade tip). As the solver is based on unstructured embedded grid methods, grids are treated as unstructured during real computation.

Figure 9: Geometry of tilt rotor.

Flow Fields of Tilt Rotor in Conversion Mode

The operating condition in this section is as follows: pitch angle is 20° and free stream velocity is 30 m/s and 50 m/s. Due to the great challenges to the unsteady flow field simulation of tilt rotor aircraft for continuous tilt process, a quasi-steady method is used here and only flow fields at four different tilt angles are calculated. Four spanwise locations are selected for analyzing the chord wise pressure variation. These locations are situated at $0.3R$, $0.5R$, $0.7R$ and $0.85R$ from the rotor center and noted as Position 1, Position 2, Position 3 and Position 4 respectively, as shown in Fig. 11.

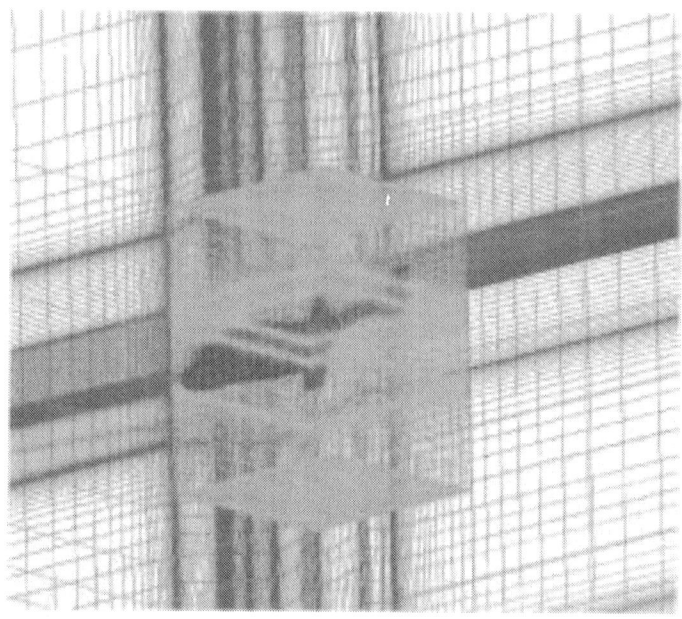

Figure 10: Tilt rotor grid system.

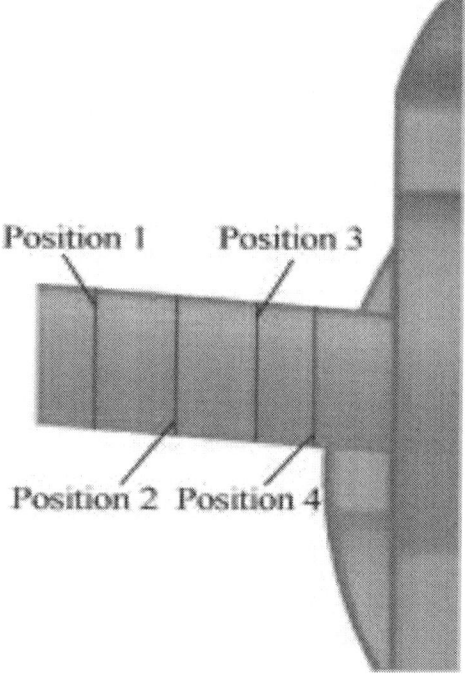

Figure 11:: Spanwise stations on the wing.

As we know, down wash induced by rotor decreases the actual angle of wing sections. Pressure distributions on the above wing locations at different rotor tilt angles are presented in Fig. 12, the symbol z/c on horizontal axis represents the nondimensional airfoil chord length. It can be seen from the figure that the pressure difference among the upper and lower airfoil sections grows up with the tilt angle increasing, which means the wake effect is weakening with the tilt angle increasing. However, there is an exception that no obvious regularity is found at Position 1, as it locates near the conjunction of wing and nacelle, which may lead to more complicated flow phenomenon.

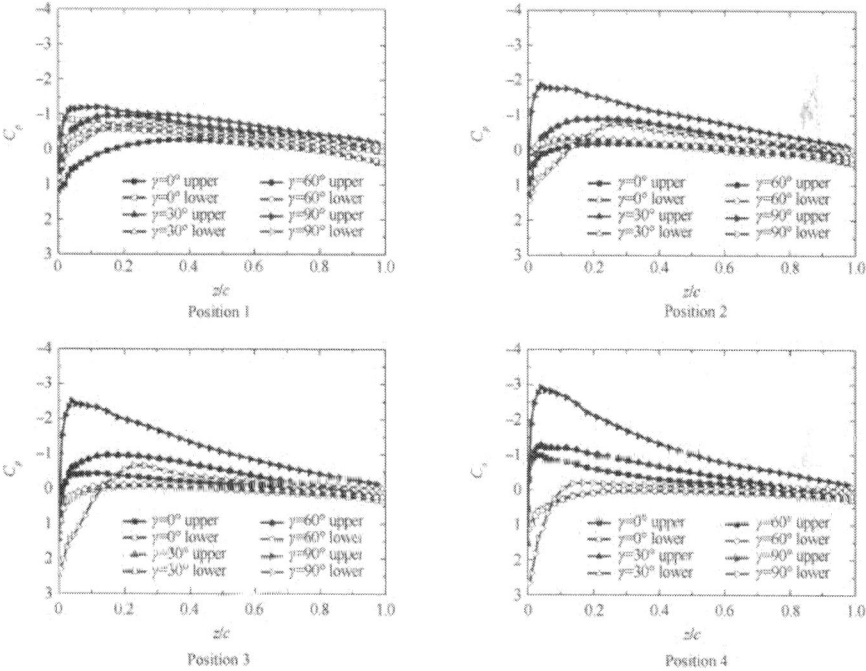

Figure 12: Pressure distributions on wing locations at different tilt angles.

Fig. 13 demonstrates pressure distribution on wing Position 3 at different free stream velocities, V. It can be seen from this figure that both the negative pressure on the upper surface and the positive pressure on the lower surface decrease with the free stream velocity increasing at the same tilt angle but the changes on the upper surface are much more obvious which leads to the increase of lift. However, there is an exception when the tilt angle is 90°, the negative pressure on the upper surface increases with the free stream velocity increasing which contributes to the decrease of lift.

In conclusion, the interaction between rotor wake and other components is complicated which relates to many factors and it changes the aerodynamic characteristic of components behind the rotor.

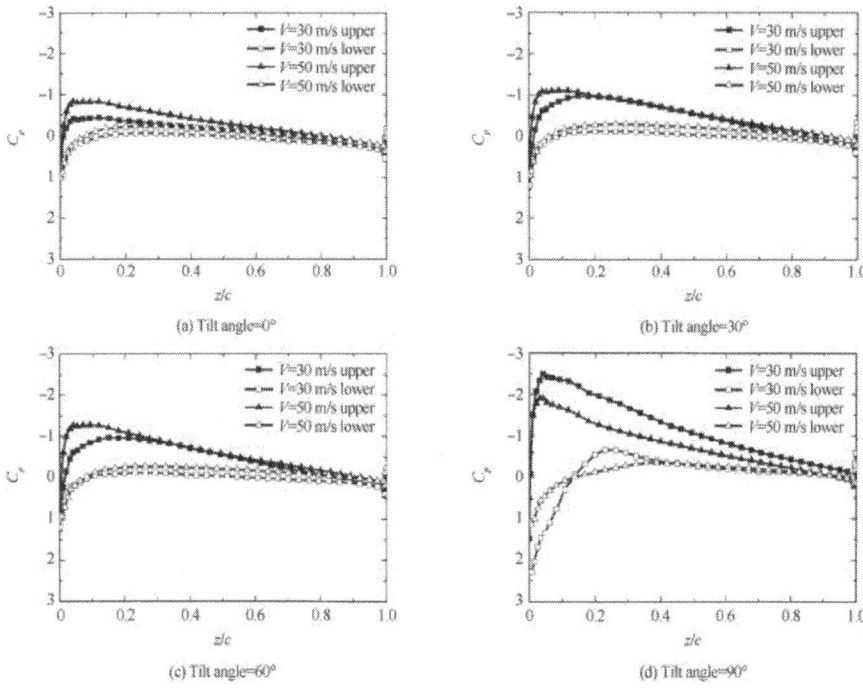

Figure 13: Pressure distributions on wing position3 at different free stream velocities.

In order to better elucidate the aerodynamic interaction between rotor and wing, the distribution of rotor thrust is analyzed, as shown in Fig. 14. Fig. 14 shows the rotor thrust coefficient distribution, with the solid line showing the result of wing/rotor model and the dashed line demonstrating that of isolated rotor model, the variables γ and C_T on the horizontal axis and vertical axis denote tilt angle and rotor thrust coefficient. It can be seen from the figure that rotor thrust decreases with the tilt angle increasing in both cases and rotor thrust is larger in the former case due to the blockage of wing to rotor wake. There is a difference about 5% when the tilt angle is 0° and the discrepancy appears a declining tendency with the tilt angle increasing.

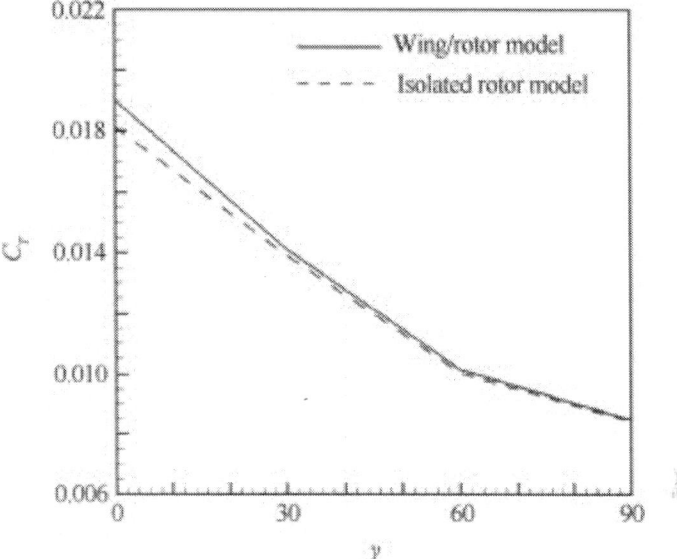

Figure 14: Rotor thrust coefficient distribution.

Fig. 15 demonstrate pressure fields and streamlines distributions on typical wing section, which is situated at $0.7R$ from the rotor center, at different rotor tilt angles for the above two free stream velocities respectively. It can be concluded that inflow is deflected and accelerated through the actuator disk and deflection effect is weakening with the tilt angle increasing. In addition, it can also be seen by comparing these two figures that the inflow is less deflected by actuator disk with the increase of free stream velocity at the same tilt angle, which is particularly obvious when tilt angle is 0°.

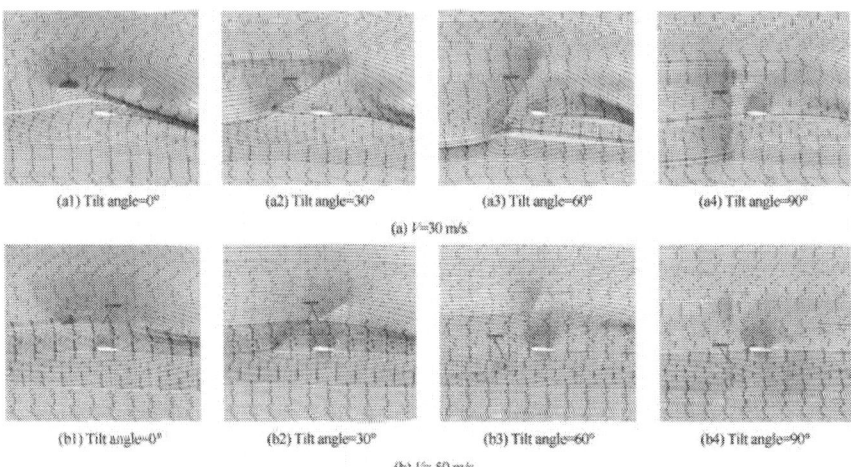

Figure 15: Pressure fields and distributions of streamlines on typical wing section at different tilt angles.

Fig. 16 shows pressure contour on fuselage and actuator surface. It can be seen from this figure that the flow induced by rotors impacts the upper wing and produces a large area of high pressure, and the influence range changes with the tilt angle variation.

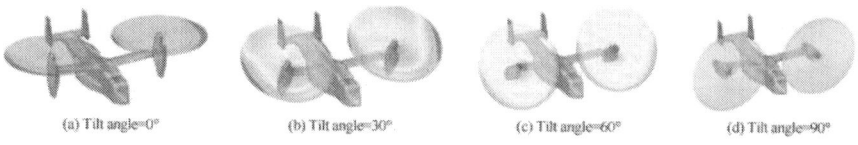

(a) Tilt angle=0°   (b) Tilt angle=30°   (c) Tilt angle=60°   (d) Tilt angle=90°

Figure 16: Pressure contour on fuselage and actuator surface at different tilt angles.

Fig. 17 illustrates streamline distributions on typical section (perpendicular to z axis and through the rotor rotation center). It can be seen from the figure that the downwash induced by rotor is blocked by wing and partial flow runs inward along the wing span then concentrates near the region of symmetry plane of fuselage and eventually deflects upward, giving birth to the rotor fountain flow, as the forward flight speed is not very high. The fountain flow phenomenon disperses with tilt angle increasing, as the rotor wake direction deflection gets weakened and less blocked by the wing. The flow structure is similar when the free stream velocity is 50 m/s and no more detailed figures are given here.

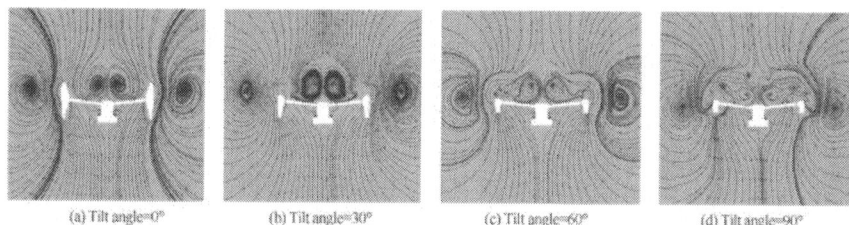

(a) Tilt angle=0°   (b) Tilt angle=30°   (c) Tilt angle=60°   (d) Tilt angle=90°

Figure 17: Streamline distributions on typical section of tilt rotor at different tilt angles ($V = 30$ m/s).

CONCLUSIONS

(1) The developed CFD solver is able to capture essential flow characteristics and predict aerodynamic interactions of tilt rotor aircraft in conversion mode.

(2) The transformation relation between different coordinate systems is deduced which successfully extends the utility of actuator disk model to rapid flow field calculation at different tilt angles.

(3) The new method for 'hole cutting' and 'donor searching' proposed is effective for grid assembling and flow fields simulation of tilt rotor in conversion mode.

ACKNOWLEDGEMENT

I would like to express my appreciation to my colleague Sun Xiaoling for her help with the modification of English expression in the manuscript.

REFERENCES

1. Johnson W, Yamauchi GK, Derby MR, Wadcock AJ. Wind tunnel measurements and calculations of aerodynamic interactions between tiltrotor aircraft. Reston: AIAA; 2003. Report No.: AIAA-2003-47.

2. Matuska D, Dale A, Lorber P. Wind tunnel test of a variablediameter tiltrotor (VDTR) model. Washington, D.C: NASA; 1994 Jan. Report No.: NASA CR177629.

3. Darabi A, Stalker A, McVeigh M, Wygnanski I. The rotor wake above a tiltrotor airplane-model in hover. Reston: AIAA; 2003 Jun. Report No.: AIAA-2003-35963.

4. Meakin RL. Unsteady simulation of the viscous flow about a V-22 rotor and wing in hover. Reston: AIAA; 1995. Report No.: AIAA-95-3463-CP.

5. Tai TC. Simulation and analysis of V-22 tiltrotor aircraft forward flight flowfield. J Aircraft 1996;33(2):369–76.

6. Gupta V, Baeder JD. Investigation of quad tiltrotor aerodynamics in forward flight using CFD. Reston: AIAA; 2002. Report No.: AIAA-2002-2812.

7. Lee-Rausch EM, Biedron RT. Simulation of an isolated tiltrotor in hover with an unstructured overset-grid RANS solver. American helicopter society 65th annual forum; 2009 May 27–29; Grapevine, TX. 2009.p. 1–19.

8. Potsdam MA, Strawn RC. CFD simulation of tiltrotor configurations in hover. American helicopter society 58th annual forum; 2002 Jun 11–13; Montreal, Canada, 2002.

9. Gupta V. Quad tilt rotor simulations in helicopter mode using computational fluid dynamics. Maryland: University of Maryland; 2005 [dissertation].

10. Liu JX. Unsteady CFD simulations of moving flow-control valves by an unstructured overset grid method. West Lafayette: Purdue University; 2008 [dissertation].

11. Lee Y, Baeder JD. Implicit hole cutting – A new approach to overset grid connectivity. Reston: AIAA; 2003. Report No.: AIAA-2003-4128.

12. Wang B, Zhao QJ, Xu G. A new moving-embedded grid method for numerical simulation of unsteady flow-field of helicopter rotor in forward flight. Acta Aerodyn Sin 2013;30(1):14–21 [Chinese].

13. Sitaraman J, Floros M, Wissink A, Potsdam M. Parallel domain connectivity algorithm for unsteady flow computations using overlapping and adaptive grids. J Comput Phys 2010;229(12): 4703–23.

14. Madrane A, Heinrich R, Gerhold T. Implementation of the embedded method in unstructured hybrid DLR finite volume Taucode. 6th overset composite grid and solution technology symposium; 2002 Oct.; Walton Beach, Florida, USA. 2002.

15. Bonet J, Peraire J. An alternate digital tree (ADT) algorithm for 3D geometric searching and intersection problems. Int J Numer Methods Eng 1991;31:1–17.

16. Rajagopalan RG, Mathur SR. Three dimensional analysis of a rotor in forward flight. Proceedings of AIAA 20th fluid dynamics, plasma dynamics and lasers conference; 1989 Jun 12–14; New York, USA. 1989. p. 1–12.

17. Roe PL. Approximate Riemann solvers, parameter vectors, and difference schemes. J Comput Phys 1981;43(2):357–72.

18. Venkatakrishnan V. On the accuracy of limiters and convergence to steady state solutions. Reston: AIAA; 1993. Report No.: AIAA-1993-0880.

19. Luo H, Baum JD. A fast, matrix-free implicit method for computing low Mach number flows on unstructured grids. Reston: AIAA; 1999. Report No.: AIAA-1999-3315.

20. McKee JW, Naeseth RL. Experimental investigation of the drag of flat plates and cylinders in the slipstream of a hovering rotor. Washington, D.C.: Langley Aeronautical Laboratory; 1958 Apr. Report No.: NACA-TN-4239.

CITATION

Ying Zhang, Liang Ye, Shuo Yang, Numerical study on flow fields and aerodynamics of tilt rotor aircraft in conversion mode based on embedded grid and actuator model, Chinese Journal of Aeronautics, Volume 28, Issue 1, February 2015, Pages 93-102, ISSN 1000-9361, http://dx.doi.org/10.1016/j.cja.2014.12.028.

CHAPTER 7

Robust Optimization of Aircraft Weapon Delivery Trajectory Using Probability Collectives and Meta-Modeling

Wang Nan[a,b], Shen Linchen [a] , Liu Hongfu[a] , Chen Jing[a] , Hu Tianjiang[a,b]

a College of Mechatronic Engineering and Automation, National University of Defense Technology, Changsha 410073, China
b Science and Technology on Aerospace Flight Dynamics Laboratory, Beijing 100094, China

ABSTRACT

Conventional trajectory optimization techniques have been challenged by their inability to handle threats with irregular shapes and the tendency to be sensitive to control variations of aircraft. Aiming to overcome these difficulties, this paper presents an alternative approach for trajectory optimization, where the problem is formulated into a parametric optimization of the maneuver variables under a tactics template framework. To reduce the size of the problem, global sensitivity analysis (GSA) is performed to identify the less-influential maneuver variables. The probability collectives (PC) algorithm, which is well-suited to discrete and discontinuous optimization, is applied to solve the trajectory optimization problem. The robustness of the trajectory is assessed through multiple sampling around the chosen values of the maneuver variables. Meta-models based on radius basis function (RBF) are created for evaluations of the means and deviations of the problem objectives and constraints. To guarantee the approximation accuracy, the meta-models are adaptively updated during optimization. The proposed approach is demonstrated on a typical air-ground attack mission scenario. Results reveal that the proposed approach is capable of generating robust and optimal trajectories with both accuracy and efficiency.

INTRODUCTION

Nowadays air-ground attack has become the main type of airborne tasks for military aircraft. Given current air-defense capabilities, the weapon delivery task is challenging even though there are no interceptors in the target area. Before engaging well-protected ground targets, deliberate plans are needed to ensure the survivability of the combat aircraft as well as the success of the attack.

In early years, work of weapon delivery planning (WDP) is mainly considered as an issue of path planning, in which a flight path is generated to guide an aircraft to reach a weapon delivery point. However, this approach is too coarse to satisfy aircraft dynamic capabilities and weapon specified constraints. With development of trajectory optimization techniques, aerodynamic models have been introduced to solve aircraft motions under a set of position and attitude constraints. Betts[1] gives a comprehensive review of the numerical methods for trajectory optimization, which formulate trajectory optimization into an optimal control problem with constraints, and summarizes two types of the representative methods, namely, direct and indirect. Indirect methods try to satisfy optimality necessary conditions derived from the application of the Pontryagin's maximum principle. Direct methods[2, 3 and 4] are based on the discretization of state and input variable sets to convert a functional problem into a nonlinear programming (NLP) problem, also called direct collocation with nonlinear programming (DCNLP), in which pseudospectral methods are main representatives, such as Guass pseudospectral, Chebyshev pseudospectral, Legendre pseudospectral. Gong et al.[2] argue that the optimal control framework is the most natural formulation for solving motion planning problems. In aircraft applications, trajectories are generated to optimize fuel and time,[5 and 6] minimize detection time during a mission under threat environment,[7, 8 and 9] and avoid obstacles or fly in constrained airspace.[10 and 11]

Although various methods have been applied to aircraft trajectory optimization, the WDP problem reveals the need for developing robust and efficient approaches to generate optimal trajectories for practical applications under complex battle environments. Firstly, terrain mask analysis is generally required in the WDP problem to find out weak points of ground threats. However, current optimal control based methods are not able to model threats with terrain mask, since explicit expressions do not exist for their irregular shapes. Secondly, planned aircraft controls may not be exactly followed during execution, but optimal control based trajectories often show a tendency to be sensitive to variations in aircraft controls and may cause failures such as a ground crash or insufficient maneuver space. To deal with the above difficulties, this paper presents a novel approach for trajectory optimization, aiming to provide an applicable alternative to conventional methods.

In this study, a new framework named tactics template is proposed to model an aircraft weapon delivery trajectory through templates of basic flight maneuvers (BFMs). The BFMs are a set of parameterized simple maneuvers, such as level turn, climb, and so on. The combination of the BFMs (also called BFM sequence) can build up complex aircraft maneuvers.[12, 13, 14 and 15] A template represents a specified BFM sequence containing a set of maneuver variables, through which the profile of the trajectory can be determined. Under this modeling, the WDP problem can be turned into a parametric optimization of the maneuver variables with independent trajectory evaluation. Besides, a detailed global sensitivity analysis (GSA) is performed to investigate the influence of the maneuver variables on the trajectory variation. The GSA results can serve as a useful guide in the identification of non-influential variables, and reduction of the optimization problem size. This can be considered as an innovative type of parametric study on trajectory optimization that has not yet been explored in the literature. The well-tailored problem is then solved by an advanced numerical solution technique named probability collectives (PC).

The PC algorithm is a newly developed multi agent system (MAS) based optimization algorithm, which has deep connections with game theory, statistical physics, and optimization.[16] The main characteristic of PC is that it operates on probability distributions of the variables rather than their values, thus it does well in handling problems with discrete or mixed type of variables. PC has been successfully applied to various benchmark and real-world optimization problems.[17, 18, 19, 20 and 21] A more general description of the advantages of PC is provided in Ref.[22] Comparison experiments also show that PC-based algorithms outperform genetic algorithms (GAs) in rate of convergence, trapping in false minima, and long term stability.[23] These features increase the flexibility of PC in the WDP problem which has a large amount of discrete variables and discontinuous objectives and constraints.

Since the goal of this paper is to create trajectories that are optimal both in the sense of mean performance and minimum variability, the robustness of the trajectory is assessed through multiple sampling and evaluation of the objectives and constraints around the chosen values of the maneuver variables. Due to the computational complexity of the trajectory evaluations, meta-models are created for both the mean value and the standard deviation of the response. The meta-models are built from carefully selected training sets generated through the design of experiments (DOE). The mean and standard deviation of the response are approximated over the variable space using radial basis function (RBF) based meta-models. To improve the global approximation accuracy, the meta-models are iteratively updated during optimization. This procedure together with the PC algorithm results in a robust and efficient trajectory optimization approach in solving the WDP problem.

PROBLEM FORMULATION

The WDP problem can be viewed as how to deliver bombs/missiles against ground targets. When preparing delivery plans, pilots used to specify delivery tactics before detailed planning,[24] which serves as guidelines to delivery trajectories. To minimize the exposure of the aircraft during delivery, the pop-up (PU) tactics is widely used to engage well-protected targets, which is studied in this paper, as shown in Fig. 1. When the tactics is selected, detailed planning then starts for further refinement of the delivery trajectories.

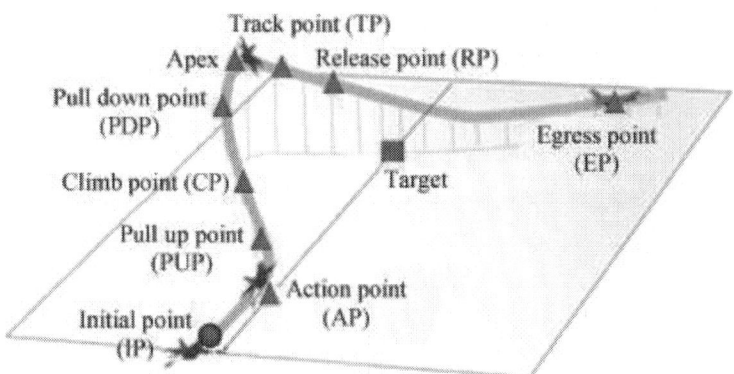

Figure1: Trajectory outline of a pop-up delivery.

Tactics Template

The tactics-oriented planning procedure has led to the development of a hierarchical framework for modeling aircraft trajectories, which is named tactics template. The tactics template contains three levels: tactics selection, maneuver optimization, and trajectory generation. A graphical illustration of the framework is given in Fig. 2. The tactics level selects the delivery tactics from the tactics library, which is represented by a series of key points specified by the tactics models (see Fig. 1). These key points divide the trajectory into sequence of simple maneuvers, which are called BFMs. There are six BFMs defined and used in this paper, including level fly, level turn, roll turn, climb, dive, and flatten out. The BFM sequence is refined at the maneuver level.

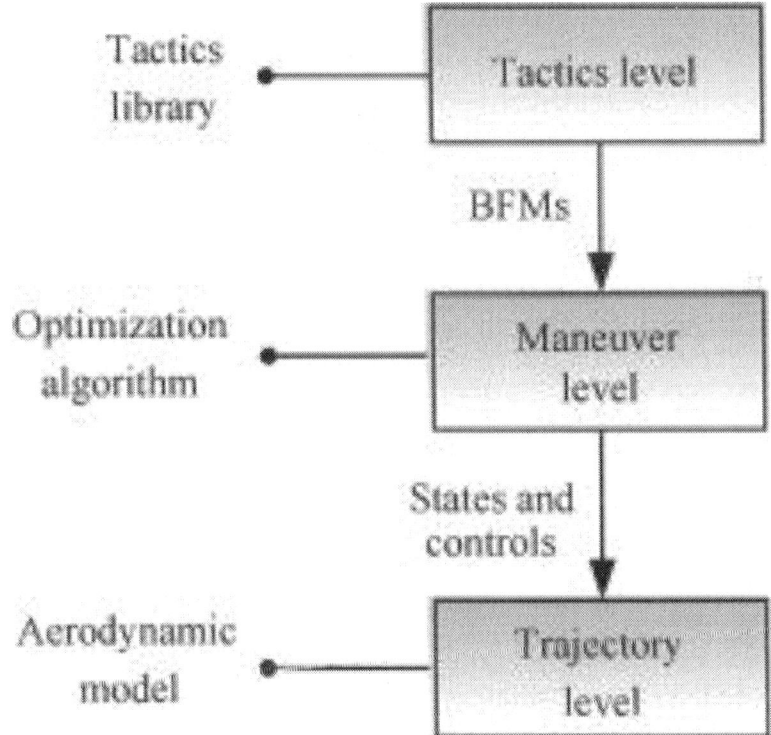

Figure2: Structure of the tactics template.

The model of BFM contains a number of steady controls and terminal conditions defined as follows:

$$BFM_j = \{c_j, y_j(t_f)\} \tag{1}$$

where j is the indexed number of the BFM, t_f the end time of the maneuver; c_j and y_j denote the control and state variables respectively. Since c_j remain unchanged during the maneuver, the dynamics of **BFM**$_j$ can be represented by:

$$y_j(t) = y_j(t_0) + \int_{t_0}^{t} f_d\big(y_j(t), c_j\big)\,dt \tag{2}$$

where $y_j(t)$ denotes the states of **BFM** at time t; t_0 is the start time of **BFM**$_j$; $f_d(\bullet)$ stands for the differential equations of the aircraft aerodynamic models. The BFMs are connected end-to-end to make up a chain representing the dynamics of the delivery tactics, as shown in Fig. 3. Though the whole variable set of the BFM chain may be large, only a few of them are closely related to the delivery tactics, which are called maneuver variables. The maneuver variables of PU are listed in Table 1. Using the formulas in Ref. [20], the other variables (state variables) of the BFMs can be calculated by the maneuver variables. It should be noted that the bombing range (B^r) is calculated by the ballistic models of the weapon used and is not included in the templates. When all the variables of the BFMs are determined, the delivery trajectory is calculated using the aircraft aerodynamic models at the trajectory level. To characterize the significance of the maneuver variables, define X^v be the maneuver variable space and let $x^v \in R^v$ be the maneuver variables. Then the original problem is transformed into a parametric optimization problem under the tactics template framework:

$$
\begin{aligned}
\min \quad & F(\phi(x^v)) \\
\text{s.t.} \quad & C(\phi(x^v)) \leqslant 0
\end{aligned}
$$

(3)

where F is the problem objective and C the constraints imposed on the trajectory; $\phi : R^v \to Y$ stands for the function that maps the maneuver variables into the states of the delivery trajectory. The advantage of this formulation is that it enables separate trajectory calculation and evaluation and allows for objectives and constraints which do not have explicit expressions as required by the optimal control mechanism.

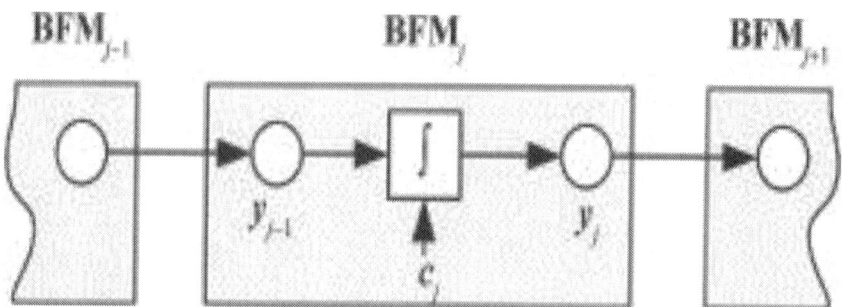

Figure 3: Illustration of the BFM chain.

Table 1: Maneuver variables of BFMs for PU.

BFM	Maneuver variable	Abbreviation
Level fly	Ingress_altitude/m	I^a
	Ingress_speed/(km·h^{-1})	I^s
	Ingress_range/km	I^r
	Ingress_azimuth/(°)	I^z
Level turn	Load_factor/G	L^f
Climb	/	/
Roll turn	Angle_off/(°)	R^o
Dive	Tracking_time/s	D^t
	Pitch_angle/(°)	D^p
	Release_altitude/m	D^a
	Release_speed/(km·h^{-1})	D^s
Flatten out	Egress_alt/m	E^a

Aircraft Aerodynamic Model

The aircraft trajectory is represented by discrete states at time points (denoted by $p_i \in P$) called nodes. Since trajectory control is the main concern of the aircraft motions in the weapon delivery task, the 3-DOF aerodynamic model [25] is adopted to calculate the states of the nodes. Assuming that there is no wind in the target area, the motions of the aircraft can be represented by the particle kinematics and dynamics equations in the geodetic coordinate system and the trajectory coordinate system respectively, as shown in Eqs. (4) and (5):

$$
\begin{cases}
\dot{x} = v \cos \gamma \cos \omega \\
\dot{z} = v \cos \gamma \sin \omega \\
\dot{h} = v \sin \gamma
\end{cases}
\tag{4}
$$

$$
\begin{cases}
\dot{v} = \dfrac{T(v) \cos \alpha - D}{m} - g \sin \gamma \\
\dot{\gamma} = \dfrac{g}{v} \left(n^d \cos \mu - \cos \gamma \right) \\
\dot{\psi} = \dfrac{g}{v \cos \gamma} n^d \sin \mu
\end{cases}
\tag{5}
$$

where x, z and h define the positions of the aircraft; v is the true airspeed; γ, ω and μ are the pitch, azimuth, and roll angles; α is the angle of attack; m the mass of the aircraft; g the acceleration of gravity, n^d the load control in the normal direction, v the accelerator control, T the thrust and D the drag forces of the aircraft. The differential equations of Eqs. (4) and (5) are solved by the Runge–Kutta solvers of MATLAB 7.6 with a uniform time step of one second. The resultant states of the nodes on the trajectory are used for further evaluations.

Trajectory Evaluation

The WDP problem contains five issues that need to be optimized: ballistic errors, threat extent, target-tracking performance, maneuverability, and safe spacing. These issues constitute the objectives and constraints of the WDP problem.

The objectives of the WDP problem include ballistic error, threat extent, and the target-tracking performance, as denoted by F^e, F^s, and F^t.

Ballistic Calculation

The ballistic calculation relies on the ballistic models of the weapon in use. In this paper, the conventional unguided bomb is studied. The B^r in Section 2.1 is calculated through simulation of the ballistic model under the release parameters (D^a, R^p, D^s). The ballistic model of the unguided bomb is given by:

$$\begin{cases} \dot{v}^b = \dfrac{-D^b}{m^b} - g \sin \gamma^b \\ \dot{\gamma}^b = -\dfrac{g \cos \gamma^b}{v^b} \end{cases} \tag{6}$$

where v^b and γ^b are the velocity and pitch angle of the bomb; D^b is the drag force of the bomb, and m^b the mass of the bomb. The ballistic error F^e is attained by looking up the statistical table of bombing accuracy of the weapon under the specified release parameters.

Threat Analysis

The threat analysis model uses the states of the nodes of the trajectory to evaluate the aircraft's exposure to the surrounding threats. The nodes which enter the threat envelope and have line-of-sight (LOS) to the threats are considered "exposed". Since the threats in the target area are mainly A-A guns and short range missiles, the detection and engagement envelopes are not distinguished. The terrain mask of the threats is pre-calculated using a fast algorithm in Ref.[26]. Since the nodes on the trajectory are of one second time spacing, F^s is calculated by summing up the exposed nodes as:

$$T^e = \sum_i \delta_s(p_i)$$

(7)

$$F^s = T^e / T^r$$

(8)

where δ_s equals 1 when p_i is exposed and 0 else; T^e is the threat exposure time of the delivery trajectory, and T^r the reaction time for the threats to engage the exposed aircraft, which is set to 10 s in this paper.

Target-tracking Evaluation

The target-tracking performance means the probability of detecting and recognizing the target with the airborne sensors during the delivery maneuvering. For electronic optical (EO) sensors, its imaging quality is represented by the resolvable cycles η across the target in slope range R, which is calculated as follows: [27]

$$\eta = L^b S^e f^c / (b^c R)$$

(9)

where L^b is the resolving line number, S^e the target equivalent dimension, f^c the focus of the camera, and b^c the height of the image surface. For different targets, η^M represents the minimum resolvable cycle required to detect and recognize the target with a probability of 50%, which is relevant to the target type and its equivalent dimension. F^t is then calculated as follows:

$$q_i^d = \frac{(\eta/\eta^M)^{2.7+0.7(\eta/\eta^M)}}{1+(\eta/\eta^M)^{2.7+0.7(\eta/\eta^M)}} \delta_r(p_i)$$

(10)

$$F^t = \sum_i (1 - q_i^d)$$

(11)

where δ_r equals 1 when the target is in LOS with the aircraft and 0 else; q_i^d denotes the instantaneous probability of target recognition at node p_i on the delivery trajectory.

Constraints

Besides the objectives, the aircraft should satisfy the maneuverability and safe spacing constraints during delivery; otherwise, a ground crash may occur. The maneuverability constraint is denoted by C^m, which is imposed on the state variables of the BFMs, e.g., $C^m = 10$ is assigned to trajectories with illegal state variables. The safe spacing constraint is denoted by C^s, which is calculated by summing up the number of nodes of the delivery trajectory which are too close to the terrain:

$$C^s = \sum_i \delta^c(p_i)$$

(12)

$$\delta^c(p_i) = \begin{cases} 1 & h(p_i) < h^t(p_i) + \Delta h^s \\ 0 & h(p_i) \geqslant h^t(p_i) + \Delta h^s \end{cases}$$

(13)

where $h(p_i)$ stands for the altitude of p_i above the sea level, $h^t(p_i)$ the elevation of the terrain at the position of p_i, and Δh^s the safe spacing. $C^m \leqslant 0$ and $C^s \leqslant 0$ mean that the constraints are satisfied along the delivery trajectory. Note that Δh^s is dependent on the pitch angle, velocity and load of the aircraft, and particularly, can be attained using a predefined safety-space table of the specified aircraft.

SOLUTION STRATEGY

The following sections present the theories and algorithms that constitute the approach for solving the WDP problem.

Global Sensitivity Analysis

Sensitivity analysis is a general concept which aims to quantify variations of an output parameter of a system regarding to changes of some input parameters. Though there has been a lot research on sensitivity analysis, most of the work concentrates on the effect of a single parameter change while the other parameters are evaluated at fixed values, which is not helpful in understanding the sensitivity behavior of a problem over the entire domain of parameter space. In order to determine the influential input parameters over defined parameter space, the GSA should be performed.

In this paper, the global sensitivity of the maneuver variables is studied. Since the problem objectives and constraints mainly depend on the delivery trajectory, the variations of the trajectory are evaluated. Let $\{p_i\}$ denote the nodes on the trajectory, the variations are calculated by summing up the distance between $\{p_i\}$ and the nodes on the "mean trajectory" calculated by the medial values of the maneuver variables:

$$P^v = \sum_i \left\| p_i - p_i^m \right\|$$

(14)

where p_i^m denotes the nodes on the mean trajectory. When the lengths of two trajectories are not the same, the shorter one is filled up with its last node. Based on the Sobol method, [28] the variance of P^v is partitioned as follows:

$$V(P^v) = \sum_{i=1}^{k} V_i + \sum_{1 \leqslant i < j \leqslant k} V_{ij} + \cdots + V_{1,2,\cdots,k}$$

(15)

where $V(P^v)$ is the total variance of P^v; $V_i = V(E(P^v | x_i^v))$ measures the main effect of the variable x_i^v; and the other terms measure the interaction effects. Eq. (13) is used to derive the sensitivity indices defined by:

$$S_i^v = \frac{V_i}{V(P^v)}$$

(16)

This index represents the main effect of x_i^v on P^v and measures the variance reduction that would be achieved by fixing that variable. Thus the values of S_i^v can be used to provide a mean to rank maneuver variables' importance on the basis of contribution to the variance of P^v, which are computed using the Monte Carlo method in Ref. [29]. The results of the GSA are presented in Fig. 4. Note that the variations of ir and iz may cause the trajectory to vary too much and are not included in the sensitivity analysis. Based on the GSA results, the variable space can be reduced by assigning the non-influential variables with fewer choices of possible values, such as E^a, D^p, and I^a.

Robust Analysis
A robust solution to the problem should not only be good in terms of optimality but also has narrow dispersion of objective function against dispersion of decision variables, which means that the optimization problem in Eq. (3) can be formulated as a robust optimization problem considering the dispersion of the objectives and constraints:

$$\begin{aligned}
&\min && (\mu_F, \sigma_F) \\
&\text{s.t.} && (\mu_C, \sigma_C) \leqslant 0
\end{aligned}$$

(17)

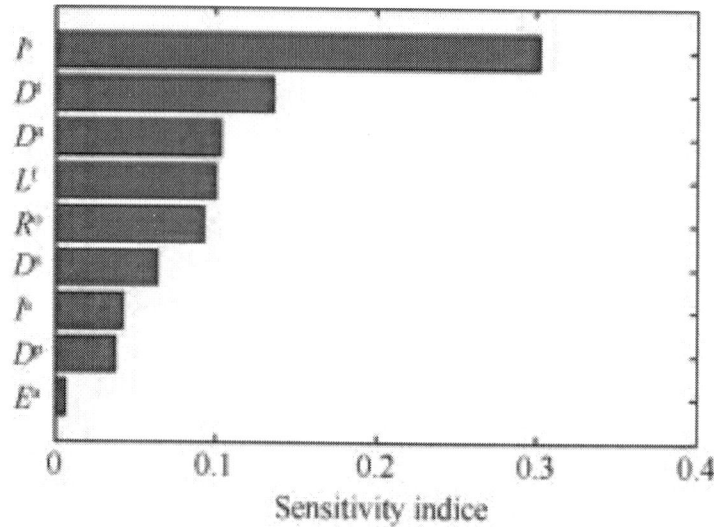

Figure 4: Sensitivity indices of maneuver variables on trajectory variation.

where μ_F and σ_F denote the mean and standard deviation of the objectives F; μ_C and σ_C denote the mean and standard deviation of the constraints C. The variance is caused by the dispersion of the maneuver variables in the tactics template. The design for six sigma (DFSS) [30] approach, which is a popular robust optimization approach in various fields of engineering, is adopted to rewrite Eq. (17) into the problem where the weighted summation of μ_F and the variance σ_F must be minimized as follows:

$$\min \quad w^\mu \mu_F + w^\sigma \sigma_F \tag{18}$$

Besides, the following inequality constraints are specified in advance to achieve the expected sigma level quality of the obtained solution:

$$\text{s.t.} \quad \mu_C \leqslant 0$$
$$\mu_C + n^S \sigma_C \leqslant U^C \tag{19}$$

where n^S denotes the sigma level and U^C the upper limits of an acceptable range of the constraints. n^S serves as a guarantee to restrict the trajectory constraints in an acceptable range under dispersions ($\pm 5\%$) of the maneuver variables. The allowed values of n^S for the six sigma level are given in Table 2. Eqs. (18) and (19) can be further converted into an unconstrained robust optimization problem through integrating the constraints into the global objective function with penalty factors:

$$\min \quad G = w^\mu \mu_F + w^\sigma \sigma_F + \lambda^\mu \mu_C +$$
$$\lambda^\sigma \max\left(n^S \sigma_C - U^C, 0\right)$$

1em(20)

(20)

Table 2: Characteristics of sigma level n^S.

Sigma level n^S	Acceptable probability/%
1	68.25
2	95.46
3	99.73
4	99.9937
5	99.999943
6	99.9999998

Since the true expressions for the mean and standard deviations of the objectives and constraints are generally not known, these entities are estimated through the Monte Carlo evaluation around each potential solution. During the evaluation, the aspects of F and C are combined together with weights that scale their importance:

$$F = w_1^F F^c + w_2^F F^s + w_3^F F^t \tag{21}$$

$$C = w_1^C C^s + w_2^C C^m \tag{22}$$

The robust optimization problem is then solved by the PC algorithm as described below.

Probability Collectives Algorithm

PC is an efficient algorithm of sampling a joint probability space of decision variables which converts an optimization problem into a convex space of probability distribution. It considers the decision variables as individual agents/players of a game being played iteratively,[17] in which a single objective called utility is to be minimized. In this paper, the utility is selected as the robust criteria defined by Eq. (20). Therefore, the optimization problem can be rewritten as:

$$\min_{\{q_i\}} \quad \int G(x^y) \prod_i q_i(x_i^y) \, dx^y$$

$$\text{s.t.} \quad \int q_i(x_i^y) \, dx_i^y = 1 \quad \forall i$$
$$q_i(x_i^y) \geqslant 0 \quad \forall i, x_i^y$$

2em(23)

(23)

where q_i denotes the probability distribution over X_i^v. It has been proven that the equilibrium of a game played by bounded rational agents is the optimizer of the Lagrangian of the probability distribution of the agents' joint-strategies, [31] in which Eq.(23) is further converted into the following terms defined as the Maxent Lagrangian: [32]

$$\min \quad L(q, T) = E_q(G) - T\sum_i S(q_i)$$
$$+ \sum_i \lambda_i \left(\int q_i(x_i)dx_i - 1 \right) \qquad 1.5em(24)$$

(24)

where $E_q(G)$ stands for the estimate of G under the joint probability distribution over X^v; $S(q_i)$ is the Shannon entropy of q_i, and T the temperature parameter. The critical points of the Maxent Lagrangian are searched through iteratively updating the probability distributions of the variables as follows: [32]

$$q_i^{k+1}(x_i^v) = q_i^k(x_i^v) - \alpha q_i^k(x_i^v)\Delta_q^k(x_i^v) \qquad (25)$$

$$\Delta_q^k(x_i^v) = \left[E_{q-i}^k(G|x_i^v) - E_q^k(G) \right] / T^k$$
$$+ S(q_i^k) + \ln q_i^k(x_i^v) \qquad 1.2em(26)$$

(26)

$$T^{k+1} = T^k T_a \qquad (27)$$

where $E_{q-i}(G|x_i^v)$ is the estimate of G given x_i^v and the joint probability distribution of all the variables except x_i^v; the superscript k denotes the kth iteration cycle; α is the step size of the gradient descent and T_a the attenuation factor that $0 < T_a < 1$. The expected value of G is evaluated using the Monte Carlo samples from the joint probability distribution of the variables as:

$$E_{q-i}^k(G|x_i^v(j)) = \frac{N_{ij}^k}{D_{ij}^k} =$$
$$\frac{\sum_m G(x_i^v(l), x_{-i}^v)\delta(x_i^v(l) - x_i^v(j)) + \kappa N_{ij}^{k-1}}{\sum_m \delta(x_i^v(l) - x_i^v(j)) + \kappa D_{ij}^{k-1}} \qquad (28)$$

where $x_i^v(j)$ denotes the j th value of x_i^v; $\delta(\cdot)$ is the Dirac delta function; κ is a data-aging factor which allows previous samples to be re-used.

A detailed illustration of the PC algorithm is presented in Table 3. It should be noted that the update rule in Eqs. (25) and (26) does not guarantee that the resultant probabilities sum to 1 and does not prevent negative probabilities. Thus the negative probabilities are set to a small positive value (such as 10^{-4}) and the probabilities are re-normalized after the update.

Table 3: Optimization procedure of PC.

Probability collectives algorithm
(1) Initialize
(a) Initialize the value space and the corresponding probability distribution of the variables. $k = 0$
(b) Set the parameters $\{T, \alpha, T_a, \kappa\}$. Set the terminal criteria
(c) Set the size of the sample block m in each iteration cycle
(2) Optimization
Repeat
(a) $k = k + 1$
(b) Jointly Monte Carlo sample the value space of the variables
(c) Evaluate the utilities for each sample
(d) Evaluate the expected utilities for each variable for each of its possible values using Eq. (28)
(e) Update the probability distribution of the variables using Eqs. (25), (26) and (27) and perform re-normalization
Until the convergence criteria is satisfied
(3) Final solution
(a) Determine the highest probability value for each variable
(b) Evaluate the objective function with this set of values

Meta-modeling

Though the PC algorithm has shown strong capability in solving discrete, discontinuous optimization problems with nonlinear objectives and constraints, the grand challenge lies in evaluating the robust criteria in which every candidate solution requires multiple calls to the computational expensive trajectory evaluation functions. To attain "good" solutions in allowable time under practical applications, it is common to use meta-models for approximating the true responses of the time-consuming evaluation functions. Though the robust optimization approach is independent of the choice of meta-models, it is advisable to choose a meta-model where the results from every evaluation could be saved and reused.

Meanwhile, the meta-model should be adaptive to the nonlinear, non-convex, and discontinuous features of the objectives and constraints of the WDP problem.

In this paper, a meta-model based on radial basis interpolation which meets the above requirements is used. The RBF interpolation is conducted on a set of training samples selected by the symmetric Latin hypercube design (SyLHD)[33] from the maneuver variable space. Assume that we are given l distinct set of maneuver variables (or called centers) where the responses are known, the RBF interpolation then takes the form:

$$\tilde{g} = \sum_{i=1}^{l} w_i^c \psi\left(\left\|x^v - x^{v(i)}\right\|\right) + p(x^v)$$

(29)

$$\psi(r) = r^2 \lg r$$

(30)

where $\tilde{g}$ is the approximated response, $\|\cdot\|$ the Euclidean norm in $\mathbf{R}^v$, $\Psi(\cdot)$ the radial basis function taking the form of thin plate spline; $x^{v(i)}$ are the RBF centers and w_i^c the corresponding weights for training; the polynomial term $p(x^v)$ is presented as a projection of G onto the space of polynomials, while the nonlinear term can be regarded as an interpolator of the residual part of the polynomial approximation. To accommodate with the robust criteria in Eq. (20), four meta-models are built for approximation of the mean and standard deviation for both the objectives and constraints, in which the outputs are denoted by $\{\tilde{\mu_F}, \tilde{\sigma_F}, \tilde{\mu_C}, \tilde{\sigma_C}\}$, respectively. Despite the output errors of the RBF interpolation are proportional to the distance between the inputs x^v and the centers $x^{v(i)}$, the global search capability is achieved by iteratively updating the RBF centers during optimization. To prevent the algorithm from prematurely converging to some possibly local optima, the constrained optimization using response surfaces (CORS)[34] mechanism is adopted to restrict the new centers to be of some distance from the previous ones. Let $x^{v(1)}, x^{v(2)}, \ldots, x^{v(l)}$ be the previously evaluated centers, the lower bound of this distance is given by:

$$\Delta_l = \max_{x^{v'} \in R^v} \min_{1 \leqslant j \leqslant l} \left\|x^{v'} - x^{v(j)}\right\|$$

(31)

where $\mathbf{x}^{v'}$ stands for the possible values of x^v. Accordingly, the new center $x^{v(k)}$ is selected by solving the auxiliary problem with the approximate objective and the dynamic distance constraint:

$$\min \qquad G^{\sim}(x^{v(k)})$$

$$\text{s.t.} \quad \left\| x^{v(k)} - x^{v(j)} \right\| \geq \varphi_k \Delta_k, \quad j = 1, 2, \cdots, l + k - 1 \qquad 0.7em(32)$$

$$(32)$$

where $0 \leq \varphi_k \leq 1$ is a parameter which balances global and local search near the current RBF centers; $G^{\sim}$ is the approximated value of G defined in Eq. (20), which is calculated by:

$$G^{\sim} = w^\mu \mu_F^{\sim} + w^\sigma \sigma_F^{\sim} + \lambda^\mu \mu_C^{\sim}$$
$$+ \lambda^\sigma \max \left(n^S \sigma_C^{\sim} - U^C, 0 \right) \qquad 0.5em(33)$$

$$(33)$$

Usually, φ_k in Eq. (32) is set in cycles. [34] In this paper, φ_k changes according to the improvement of G of the latest iteration as:

$$\varphi_k = \begin{cases} \max\left(\dfrac{\varphi_{k-1}}{\beta}, 1\right) & \Delta_k^g < \Delta^{g\min} \\[2ex] \dfrac{\varphi_{k-1}}{\beta} & \text{Otherwise} \end{cases}$$

$$(34)$$

$$\Delta_k^g = G\left(x^{r(k-1)}\right) - G\left(x^{r(k-2)}\right) \qquad (35)$$

$$\Delta^{g\min} = (G_{\min} - G_{1/4}) I_{\min} \qquad (36)$$

where β is the distance control factor $(0 < \beta < 1)$; $G_{\min}$ stands for the current optimal value of G; $G_{1/4}$ the lower quartile (1/4) value of G in history; $I_{\min}$ the improvement control factor. Under this scheme, the distance constraint enlarges as the improvement in G becomes low, forcing the algorithm to search the areas which are farther from the centers for potential optimal solutions. When φ_k exceeds 1, it is set back to an initial value φ_0. To be compatible with the PC algorithm, the auxiliary problem is further converted into the unconstrained form as:

$$\min \quad \tilde{G}(x^{v(k)}) + \lambda^\varepsilon \sum_{j=1}^{l+k-1} \left(\beta_j \Delta_k - \left\| x^{v(k)} - x^{v(j)} \right\| \right)$$

$$(37)$$

where λ^ε is the penalty factor. When the kth optimal values $x^{v(k)}$ of the variables are found, the true $G(x^{v(k)})$ is evaluated and the RBF models are updated with the new set of centers. The iteration cycle repeats until the following convergence criteria are satisfied:

$$|G - G^{\sim}| \leqslant \varepsilon_1, \quad |G^k - G^{k-1}| \leqslant \varepsilon_2 \tag{38}$$

where the G^k is the abbreviation for $G(x^{v(k)})$.

Overall Solution Procedure

The main steps of the solution procedure of the proposed approach are given in Fig. 5. More details can be found in Table 4.

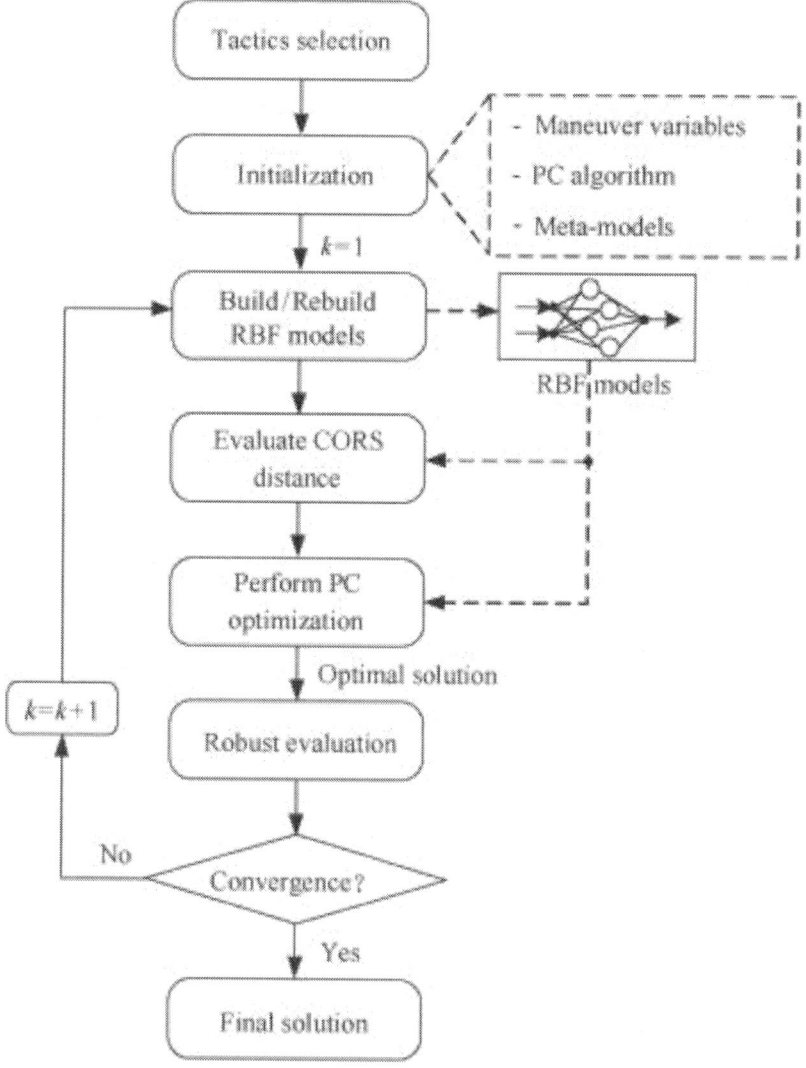

Figure5: Main steps of the solution procedure of the proposed approach.

Table 4: Robust solution procedure of the WDP problem.

Robust optimization with meta-modeling
(1) Select the delivery tactics from the tactics library
(2) Initialize the value space of the maneuver variables of the tactics template based on sensitivities from GSA
(3) Initialize the PC algorithm
(4) Initialize the meta-model
(a) Set the initial number of training centers l
(b) Generate l set of values of the maneuver variables using SyLHD
(c) Perform Monte Carlo sampling and evaluation around each set, and calculate the mean and standard deviations of the objectives and constraints
(5) Generate l^- set of values of the maneuver variables using SyLHD
(6) Optimization
(a) $k = 1$.
Repeat
(b)
Build the four RBF models of the mean and standard deviations for the objectives and constraints with the l training centers
(c) For each RBF model, evaluate Δ_k for the its centers under the l^- set of values of the maneuver variables using Eq. (31); and calculate the corresponding φ_k using Eq. (34)
(d) Perform PC optimization of the maneuver variables with the RBF models based on the utility of Eq. (37)
(e) Calculate the mean and standard deviations of the objectives and constraints the optimal solution
(f) Add the optimal solution to the training centers, $k = k + 1$
Until the convergence criteria in Eq. (38) is satisfied or $k \geqslant$ kmax
(7) Final solution
Select the latest optimal solution as the final solution to the problem

EXPERIMENT AND RESULTS

Tested Scenario

In order to demonstrate the performance of the proposed approach, a typical air-ground attack mission scenario is presented. The scenario is of 150 km × 150 km in size, which contains one target surrounded by three anti-air missile threats. A graphical view of the scenario is given in Fig. 6, where the triangle stands for the target and the outlines stand for the threat envelopes at 1000 m and 2000 m under terrain mask (the maximum engagement range of the threats is 40 km). A 3D view of the terrain of the scenario is given in Fig. 7. As shown in Fig. 6, the target area is fully covered by threat envelopes at 2000 m, so a well-designed PU delivery trajectory becomes extremely important to the survivability of the combat aircraft.

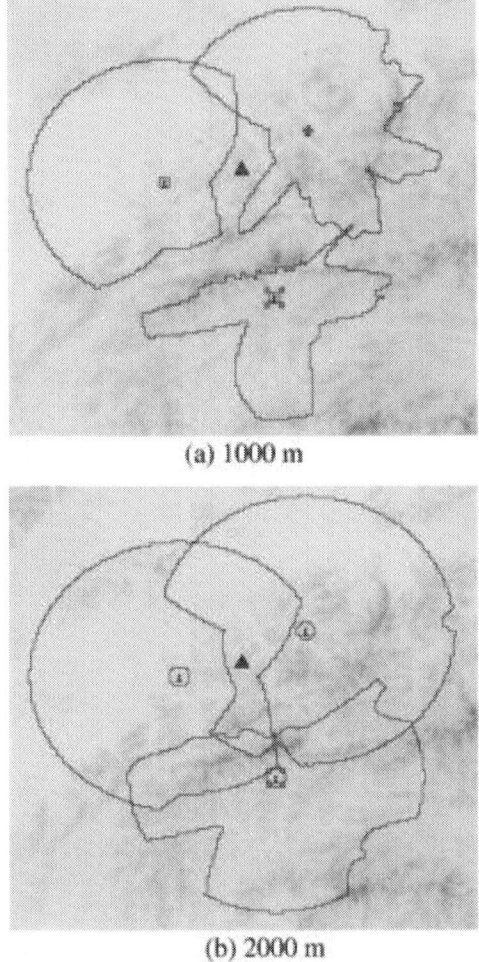

(a) 1000 m

(b) 2000 m

Figure 6: Test mission scenario displayed with threat envelopes under terrain mask.

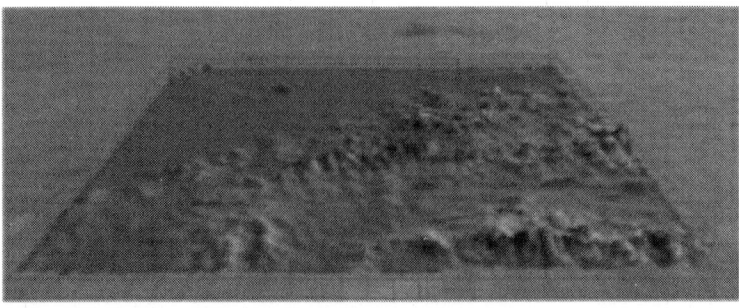

Figure 7: 3D view of the mission scenario.

Solution
Under the above mission scenario, a robust optimization of the delivery trajectory based on the proposed approach is performed. The maneuver variables are discretized by constant steps within their boundaries according to the GSA results, as shown in Table 5, with a total decision space as large as 10^{13}. Some main parameters' values of the optimization model and the PC algorithm are given in Table 6. To evaluate the influence of initial numbers of training samples on the final solutions, different sets of training samples ($l = 20, 50, 100$) for the meta-models are generated. Besides, to evaluate the robust criteria in Eq. (20), 10 samples ($l^+ = 10$) are generated for robust analysis of each potential solution.

Table 5: Maneuver variables of BFMs for PU.

Abbreviation	Boundaries	Steps
I^a	[200, 800]	50
I^s	[500, 1000]	10
I^r	[10, 15]	0.5
I^z	[0, 360)	1
L^f	[4, 9]	1
R^o	[−90, −40]∪[40, 90]	5
R^t	[4, 10]	1
R^p	[5, 60]	5
D^a	[1200, 2000]	20
D^s	[550, 1000]	50
E^a	[500, 1000]	100

Table 6: Values of the main parameters of PC and the optimization model.

Parameter	Value	Parameter	Value
T	0.1	w_1^C, w_2^C	1, 2
α	0.2	w^μ, w^σ	1, 1
T_a	0.95	$\lambda^\mu, \lambda^\sigma$	10, 10
κ	0.5	U^C	1
w_1^F, w_2^F, w_3^F	20, 1, 0.5	n^S	3

The platform for the experiment is a Core2 Duo 2.8 GHz computer installed with Windows XP. All the codes are programmed in VC++.

The robust optimal solution over 30 iterations with $l = 100$ is presented in Table 7, which is compared to a non-robust optimal solution found during optimization. The corresponding objectives and constraints of the two solutions are given in Table 8. Graphical views of the delivery trajectories of the two solutions are given in Fig. 8, in which the threat envelopes are calculated under terrain mask. As shown in the figure, there is only small distance left for the level-fly BFM in the trajectory of the non-robust optimal solution, and small changes of the maneuver variables may further reduce the distance and thus violate the maneuverability constraints. Fig. 9 displays the results of 10 Monte Carlo evaluations of the objectives and constraints of the two solutions under ±10% dispersion of the maneuver variables. It can be seen that though the objectives and constraints of the non-robust optimal solution are better than those of the robust one, it is not tolerable to small changes of the maneuver variables.

Table 7: Solutions of the maneuver variables for the tested scenario.

Abbreviation	Value (Robust)	Value (Non-robust)
I^a	800	200
I^s	750	750
I^r	15	14
I^z	220	300
L^f	7	7
R^o	−60	75
R^t	10	9
R^p	50	40
D^a	1650	1600
D^s	550	950
E^a	1000	1000

Table 8: Values of objectives and constraints of the robust and non-robust optimal solutions.

Objective and constraint	Value (Robust)	Value (Non-robust)
F^e	0.2302	0.1991
F^s	8.5	6.7
F^t	15.0162	13.038
C^m	0	0
C^s	0	0
G	23.941	237.5891

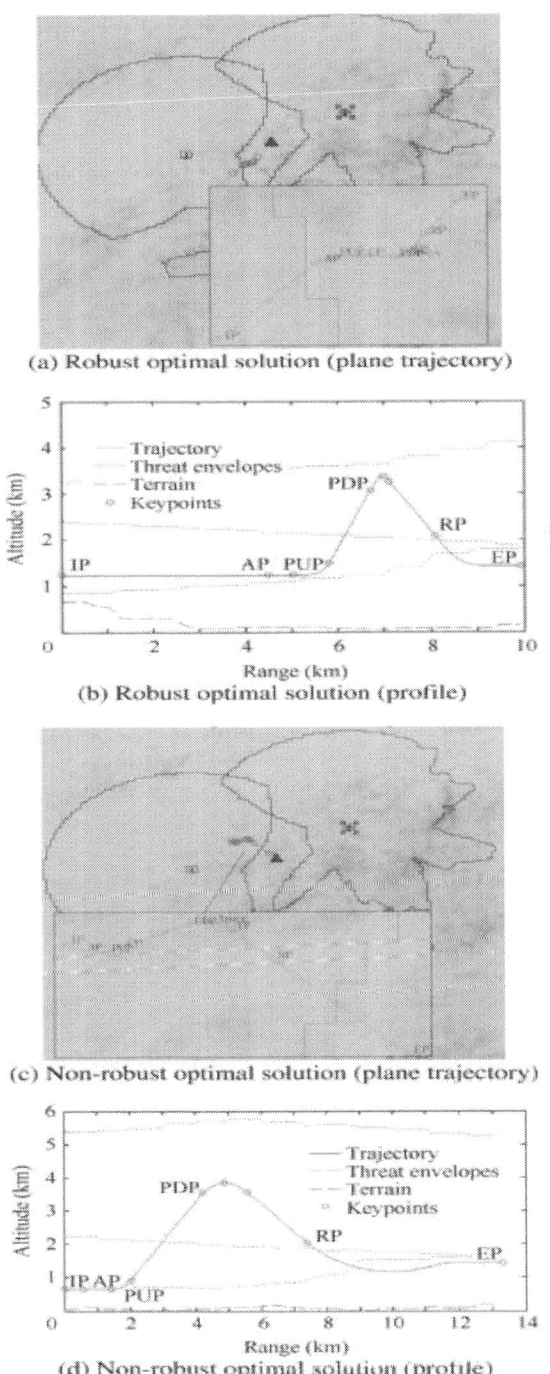

(a) Robust optimal solution (plane trajectory)

(b) Robust optimal solution (profile)

(c) Non-robust optimal solution (plane trajectory)

(d) Non-robust optimal solution (profile)

Figure8: Level projection and profiles of the trajectories of the robust and non-robust optimal solutions.

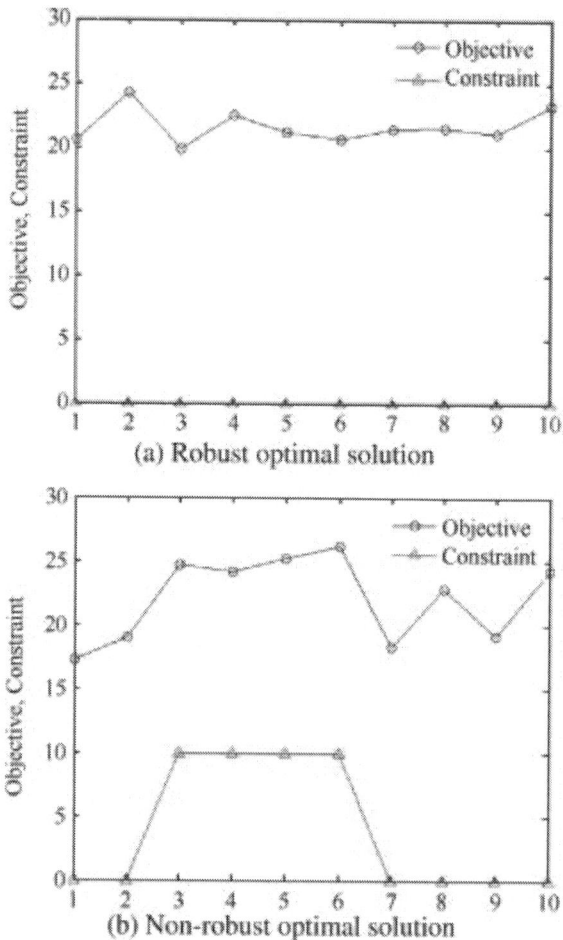

(a) Robust optimal solution

(b) Non-robust optimal solution

Figure 9: Monte Carlo evaluations of the robust and non-robust optimal solutions.

To demonstrate the efficiency of the proposed approach, the evolutions of $\mu^F, \sigma^F, \mu^C,$ and σ^C during optimization are given in Fig. 10. The optimization performance under different initial training samples ($l = 20$, 50, 100) for the meta-models is compared in the figure. It can be seen that larger initial training samples can result in improved optimization performance, but the improvement is not remarkable between $l = 50$ and $l = 100$, implying that a moderate number of initial training samples may be a better choice to balance solution accuracy and efficiency. Besides, the sharp changes of the objectives and constraints reflect the effects of CORS that restrict the new solutions to be of some distance from the previous ones. The execution time of 30 iterations is about 150–390 s according to different initial numbers of training samples.

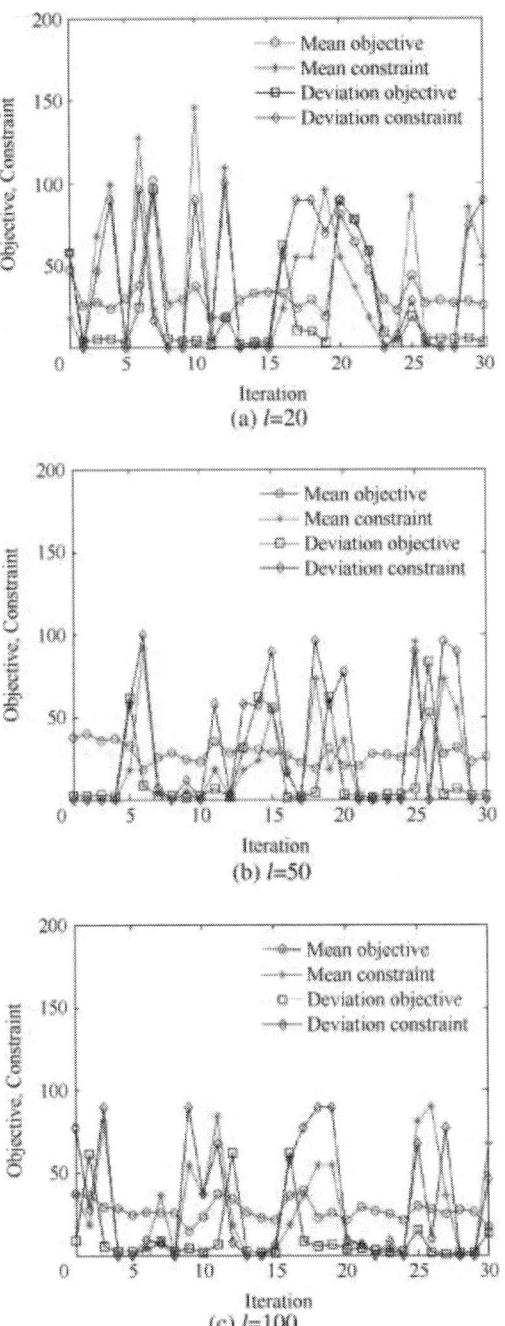

Figure 10: Evolution of the mean and standard deviation of the problem objectives and constraints. The response values shown are from actual evaluations.

Error Analysis

To check the validity of the meta-models, the root mean square (RMS) errors are studied:

$$RMS = \sqrt{\sum_{i=1}^{P} \frac{(\tilde{y_i} - y_i)^2}{P}}$$

(39)

where $\tilde{y_i}$ is the approximated response by the meta-model, y_i the actual response, and P the number of tested points. The RMS errors for both the mean values and standard deviations are tested using 50 randomly selected testing samples during optimization, as shown in Table 9. Due to the large decision space and the complexity of the problem objectives and constraints, the errors are the price for trying to maintain a global approximation. However, the RMS errors are reduced during optimization, since the meta-models are updated with well-selected samples produced by the proposed approach. As the optimization proceeds, the global approximation accuracy of the meta-models grows. It is up to the user to decide the maximum number of iterations that balances the solution accuracy and efficiency.

Table 9: RMS errors for the mean and standard deviation of the objectives and constraints during optimization ($l = 100$).

Iteraion	μ^F	σ^F	μ^C	σ^C
1	40.9101	32.4105	48.742	31.439
5	36.4445	34.2401	32.969	32.28
10	33.2296	37.3212	28.735	35.5329
15	31.0122	29.7472	26.576	27.9633
20	26.7008	19.9179	25.61	19.7096
25	26.5585	19.591	24.755	19.8522
30	25.7165	19.7204	24.893	19.8014

DISCUSSION

The main advantage of the proposed approach to robustness of the delivery trajectory is that it turns an aircraft trajectory optimization into a parametric optimization problem through the use of tactics template. As mentioned earlier, this formulation successfully avoids the shortcomings of current optimal control based trajectory optimization techniques, and makes it well suited for trajectory optimization problems in complex threat environments. Though the size of the variable space may be large, it is reduced with the GSA method. To account for the robustness of the maneuver variables, several evaluations are made around each potential solution.

The second advantage of the approach is that it successfully combines the PC-based optimization algorithm and meta-models to solve the computational expensive trajectory optimization problem. The PC algorithm requires no prior knowledge and initial guess of the optimal solutions, and is applicable to a wide range of nonlinear, discrete, and discontinuous optimization problems, including trajectory optimization with maneuver variables. To reduce the computational efforts of evaluating the complex objectives and constraints during optimization, the RBF-based meta-models are built to replace the trajectory evaluation procedure. However, in order to attain satisfied approximation accuracy, appropriate number of training samples and samples for robust analysis should be set by the user. It is advisable to study the GSA results, as they indicate the number of influential variables to the problem. It is also suggested by the experiment results that a moderate number of initial training samples is more preferable, since the decision variable space is usually too large to be fully covered.

The $\pm 10\%$ range of the dispersion of the maneuver variables is selected according to typical execution errors for manned aircraft. For unmanned aircraft, this range can be calculated based on the trajectory following capabilities of the aircraft. The sigma level n^s may also be changed according to different mission and safety requirements, since it serves as a balance between optimality and robustness.

Furthermore, due to the complexity and nonlinearity of the problem objectives and constraints, global convergence is not guaranteed. When searching for a global optimum, it is customary to try different distributions of the initial training samples for the meta-models. This is not done in this paper, but it should be considered in more comprehensive studies.

CONCLUSIONS

This paper is presented as a preliminary study of the aircraft weapon delivery trajectory optimization problem. The aim of this paper is to create a robust and efficient approach for aircraft trajectory optimization under complex mission environments. The approach works well on the tested mission scenario and the key feature of the approach is that it turns trajectory optimization into a parametric optimization problem and solves it with the well-decided numeric optimization techniques. Moreover, the approach is applicable to other trajectory optimization cases with different tactics template formulations.

ACKNOWLEDGEMENT

This study was supported by Open Research Foundation of Science and Technology on Aerospace Flight Dynamics Laboratory (No. 2012afd1010).

REFERENCES

1. Betts JT. Survey of numerical methods for trajectory optimization. J Guidance Control Dyn 1998;21(2):193–207.
2. Gong Q, Lewis LR, Ross IM. Pseudospectral motion planning for autonomous vehicles. J Guidance Control Dyn 2009;32(3):1039–45.
3. Laurent-Varin J, Bonnans F, Be´rend N, Haddo M. Interior-point approach to trajectory optimization. J Guidance Control Dyn 2007;30(5):1228–39.
4. Geiger BR, Horn JF, Sinsley GL, Ross JA, Long LN, Niessner AF. Flight testing a real-time direct collocation path planner. J Guidance Control Dyn 2008;31(6):1575–86.
5. Ringertz U. Optimal trajectory for a minimum fuel turn. J Aircr 2000;37(5):932–5.
6. Soler M, Olivares A, Staffetti E. Hybrid optimal control approach to commercial aircraft trajectory planning. J Guidance Control Dyn 2010;33(3):985–91.
7. Norsell M. Radar cross section constraints in fight-path optimization. J Aircr 2003;40(2):412–5.
8. Norsell M. Multistage trajectory optimization with radar range constraints. J Aircr 2005;42(2):849–57.
9. Inanc T, Muezzinoglu MK, Misovec K, Murray RM. Framework for low-observable trajectory generation in presence of multiple radars. J Guidance Control Dyn 2008;31(6):1740–9.
10. Mattei M, Blasi L. Smooth flight trajectory planning in the presence of no-fly zones and obstacles. J Guidance Control Dyn 2010;33(2):454–62.

11. Patel RB, Goulart PJ. Trajectory generation for aircraft avoidance maneuvers using online optimization. J Guidance Control Dyn 2011;34(1):218–30.
12. Ghosh R, Tomlin C. Nonlinear inverse dynamic control for modebased flight. In: Proceedings of AIAA Guidance, Navigation, and Control Conference and Exhibit 2000;1–11.
13. Frazzoli E, Dahleh MA, Feron E. Real-time motion planning for agile autonomous vehicles. J Guidance Control Dyn 2002;25(1):116–29.
14. Koyuncu E, Ure NK, Inalhan G. A probabilistic algorithm for mode based motion planning of agile unmanned air vehicles in complex environments. In: International Federation of Automatic Control (IFAC'08) World Congress 2008;17.
15. Ure NK, Inalhan G. Design of a multi modal control framework for agile maneuvering UCAV. In: IEEE Aerospace Conference 2009.
16. Bieniawski SR. Distributed optimization and flight control using collectives dissertation. Stanford: Stanford University; 2005.
17. Bieniawski SR, Wolpert DH, Kroo IM. Discrete, continuous, and constrained optimization using collectives. In: 10th AIAA/ISSMO Multidisciplinary Analysis and Optimization Conference 2004;1–9.
18. Kulkarni AJ, Tai K. Probability collectives: a multi-agent approach for solving combinatorial optimization problems. Appl Soft Comput 2010;10(3):759–71.
19. Kulkarni AJ, Tai K. Probability collectives: a distributed optimization approach for constrained problems. In: IEEE Congress on Evolutionary Computation (CEC) 2010;1–8.
20. Shah K, Kumar M. Resource management in wireless sensor networks using collective intelligence. In: International Conference on Intelligent Sensors, Sensor Networks and Information Processing (ISSNIP) 2008;423–8.
21. Antoine NE, Bieniawski SR, Wolpert DH, Kroo IM. Fleet assignment using collective intelligence. In: 42nd AIAA Aerospace Sciences Meeting and Exhibit 2004;1–5.
22. Kulkarni AJ, Tai K. Probability collectives: a decentralized, distributed optimization for multi-agent systems. In: Proceedings of the Online World Conference on Soft Computing in Industrial Applications 2008.
23. Huang CF, Bieniawski SR. A comparative study of probability collectives based multi-agent systems and Genetic Algorithms. In: Genetic and Evolutionary Computation Conference; 2005. p. 751–2.
24. HQ ACC/DOT. F-16 combat aircraft fundamentals; 1996. http://www.fas.org/man/dod-101/sys/ac/docs/16v5.pdf.
25. Xu MY, Ding SB. Flight dynamics. Beijing: Science Press; 2003, Chinese.
26. Leavitt CA. Real-time in-flight planning. In: Proceedings of the IEEE National Conference on Aerospace and Electronics 1996;1:83–9.

27. Leachtenauer JC, Dirggers RG. Surveillance and reconnaissance imaging systems (modeling and performance prediction). Norwood: Artech House Publishers; 2007.

28. Saltelli A, Sobol IM. Sensitivity analysis for nonlinear mathematical models: numerical experience. Math Models Comput Exp 1995;11(7):16–28.

29. Homma T, Saltelli A. Importance measures in global sensitivity analysis of nonlinear models. Reliab Eng Syst Saf 1996;52(1):1–17.

30. Shimoyama K, Oyama A, Fujii K. Development of multi-objective six-sigma approach for robust design optimization. J Aerosp Comput Inf Commun 2008;5:215–33.

31. Wolpert DH. Information theory-the bridge connecting bounded rational game theory and statistical physics. In: Braha D, Minai Y, Bar-Yam Y, editors. Complex Engineering Systems. New York: Springer; 2006. p. 262–90.

32. Wolpert DH, Strauss CEM, Rajnarayan D. Advances in distributed optimization using probability collectives. Adv Complex Syst 2006;9(4):383–436.

33. Ye KQ, Li W, Sudjianto A. Algorithmic construction of optimal symmetric latin hypercube designs. J Stat Plann Inference 2000;90(1):145–59.

34. Regis RG, Shoemaker CA. Constrained global optimization of expensive black box functions using radial basis functions. J Global Optim 2005;31(1):153–71.

CITATION

Wang Nan, Shen Lincheng, Liu Hongfu, Chen Jing, Hu Tianjiang, Robust optimization of aircraft weapon delivery trajectory using probability collectives and meta-modeling, Chinese Journal of Aeronautics, Volume 26, Issue 2, April 2013, Pages 423-434, ISSN 1000-9361, http://dx.doi.org/10.1016/j.cja.2013.02.020.

CHAPTER 8

Unsteady Aerodynamic Modeling at High Angles of Attack Using Support Vector Machines

Wang Qing, Qian Weiqi, He Kaifeng[ab]

[a] State Key Laboratory of Aerodynamics, Mianyang 621000, China
[b] Computational Aerodynamics Institute, China Aerodynamics Research and Development Center, Mianyang 621000, China

ABSTRACT

Accurate aerodynamic models are the basis of flight simulation and control law design. Mathematically modeling unsteady aerodynamics at high angles of attack bears great difficulties in model structure determination and parameter estimation due to little understanding of the flow mechanism. Support vector machines (SVMs) based on statistical learning theory provide a novel tool for nonlinear system modeling. The work presented here examines the feasibility of applying SVMs to high angle-of-attack unsteady aerodynamic modeling field. Mainly, after a review of SVMs, several issues associated with unsteady aerodynamic modeling by use of SVMs are discussed in detail, such as selection of input variables, selection of output variables and determination of SVM parameters. The least squares SVM (LS-SVM) models are set up from certain dynamic wind tunnel test data of a delta wing and an aircraft configuration, and then used to predict the aerodynamic responses in other tests. The predictions are in good agreement with the test data, which indicates the satisfying learning and generalization performance of LS-SVMs.

INTRODUCTION

Unsteady aerodynamics at high angles of attack plays an increasingly important part in modern aircraft design. In high angle-of-attack maneuvers, the flow field around the aircraft is extremely complex and the aerodynamics shows strong nonlinearity and unsteadiness. As a result, the conventional aerodynamic database, composed of static test data, dynamic derivatives and rotary-balance data, does not meet the requirements of flight simulation and control law design. The database needs to involve dynamic test data. Unfortunately, the aerodynamic characteristics at high angles of attack, especially in post-stall maneuvers, cannot be predicted simply by interpolation among limited test data, as done at pre-stall angles of attack. A feasible solution is to set up aerodynamic models, which describe the dependence of aerodynamics upon the motion history, from a certain number of static and dynamic wind tunnel test data.

Currently, the researches on high angle-of-attack unsteady aerodynamic modeling evolve in two directions: mathematic methods and artificial intelligent methods. The developed mathematic models include those in form of generalized aerodynamic derivatives,[1] nonlinear indicial response,[2, 3 and 4] internal state-space,[5 and 6] differential equations,[7, 8, 9 and 10] hybrid representation of nonlinear indicial response and internal state-space,[11] flow incidence rate,[12] etc. They are based on the understanding of physical phenomenon and mechanism. The available intelligent methods include fuzzy logic (FL)[13, 14 and 15] and neural networks (NNs),[16, 17, 18, 19 and 20] which are suitable to black-box system modeling especially. In addition, reduced order models of nonlinear and unsteady aerodynamics, based on indicial response functions, have been developed.[21, 22 and 23] These functions can be estimated via analytical, experimental or computational methods. The analytical solutions are limited only to two-dimensional airfoils in incompressible and inviscid flows.[21] Experimental tests are practically nonexistent for indicial response functions. CFD calculations have recently been used to determine indicial response functions for the given aircraft configurations.[22 and 23]

A large number of wind tunnel tests[24, 25 and 26] and CFD simulations[27, 28 and 29] have been conducted to study unsteady aerodynamics of maneuvering aircraft at high angles of attack. One has acquired some knowledge about the effects of reduced frequency, amplitude, and mean angle of attack on unsteady aerodynamics in forced-oscillations. However, many problems in the area of unsteady flow mechanism at high angles of attack, particularly in the case of multiple degree-of-freedom (DOF) coupling motions, have not been solved yet, which leads to great difficulties in model structure determination and parameter estimation when mathematically modeling unsteady aerodynamics. In this case, intelligent methods are gaining popularity. They let computers learn the available static and dynamic wind tunnel test data and then predict the aerodynamic responses of aircraft in flight. In intelligent methods, how to determine the optimal model structure is a critical problem, which has not been solved perfectly by FL and NNs. Support vector machines (SVMs), a new type of statistical learning strategy, embody the structural risk minimization (SRM) principle, which has been shown to be superior to the traditional empirical risk minimization (ERM) principle, employed by conventional FL and NNs.[30, 31 and 32] Therefore, SVMs exhibit more excellent empirical performance than FL and NNs. Another attractive feature of SVMs is the global optimality. By introducing a nonlinear map from input space to feature space, SVMs transform a nonlinear system modeling problem to a quadratic programming, which can achieve global minimum.

This paper presents a pioneer study of using SVMs to model high angle-of-attack unsteady aerodynamics. After a review of SVMs and least squares SVMs (LS-SVMs), an extension of the standard SVMs, a simulation experiment of a two-dimensional nonlinear system is performed to validate the empirical performance of SVMs. Several issues associated with application of SVM method in unsteady aerodynamic modeling field are discussed in detail, such as selection of input variables, selection of output variables, and determination of SVM parameters. The LS-SVM method is applied to aerodynamic modeling of a pitching delta wing and a rolling aircraft configuration. The satisfying learning and generalization performance exhibited by the applications indicates the feasibility of applying SVMs to high angle-of-attack unsteady aerodynamic modeling.

SVMS FOR NONLINEAR SYSTEM MODELING

SVMs, a novel tool in the area of machine learning, were first proposed by Vapnik[33] in 1995. Originally, they were developed for pattern recognition problems. Recently, they have been successfully extended to nonlinear function approximation and nonlinear system modeling.

SVM Regression

The SVMs used in system modeling are called SVM regression, or support vector regression (SVR) in short. For a problem of multi-input single-output (MISO) nonlinear system modeling, SVMs approximate the nonlinear function by a linear regression:

$$y = f(x) = w^T \varphi(x) + b \quad x \in \mathbf{R}^m, y \in \mathbf{R} \tag{1}$$

in a high-dimensional feature space F. Here $\varphi(x)$ denotes a nonlinear transformation from input space $\mathbf{R}^m$ to feature space F, w is weighting vector, and b is bias.

Suppose that a finite number set of sample data $\{(x_i, y_i), i = 1, 2, \ldots, n\}$ have been obtained by experimental measurement. If all the training data can be fitted by the function Eq. (1) with ε precision, then

$$\begin{cases} y_i - \boldsymbol{w}^{\mathrm{T}} \varphi(x_i) - b \leqslant \varepsilon \\ \boldsymbol{w}^{\mathrm{T}} \varphi(x_i) + b - y_i \leqslant \varepsilon \end{cases} \quad i = 1, 2, \cdots, n \tag{2}$$

Sometimes, however, this may not be the case, or we may also want to allow for some error. One can introduce slack variables ξ and ξ^* to cope with otherwise infeasible constrains:

$$\begin{cases} y_i - \boldsymbol{w}^{\mathrm{T}} \varphi(x_i) - b \leqslant \varepsilon + \xi_i \\ \boldsymbol{w}^{\mathrm{T}} \varphi(x_i) + b - y_i \leqslant \varepsilon + \xi_i^* \end{cases} \quad i = 1, 2, \cdots, n \tag{4}$$

The SRM principle yields the optimization goal:

$$\min_{w, b, \xi, \xi^*} J = \frac{1}{2} \|\boldsymbol{w}\|^2 + c \sum_{i=1}^{n} (\xi_i + \xi_i^*) \tag{4}$$

where c is penalty factor and a pre-specified constant determining the training error and the regression function flatness.

Using the Lagrange function method together with the dual variables to find the solution of the above problem can lead to a quadratic programming (QP) problem:

$$\begin{cases} \max_{a, a^*} J = -\frac{1}{2} \sum_{i,j=1}^{n} (a_i - a_i^*)(a_j - a_j^*) K(x_i, x_j) - \varepsilon \sum_{i=1}^{n} (a_i + a_i^*) \\ \qquad + \sum_{i=1}^{n} y_i(a_i - a_i^*) \\ \text{s.t.} \begin{cases} \sum_{i=1}^{n} y_i(a_i - a_i^*) = 0 \\ a_i, a_i^* \in [0, c] \end{cases} \end{cases} \tag{5}$$

where a_i and a_i^* are the Lagrange multipliers; $K(x_i, x_j)$ is called kernel function. Its value is equal to the inner product of two vectors x_i and x_j in the feature space $\varphi(x_i)$ and $\varphi(x_j)$, i.e., $K(x_i, x_j) = \varphi(x_i) \cdot \varphi(x_j)$.

Solving the above QP problem with inequality constrains, the Lagrange multipliers a_i and a_i^* can be determined. Then, the weighting vector w and the bias b can be derived from Karush–Kuhn–Tucker's (KKT's) condition.

$$\begin{cases} w = \sum_{i=1}^{n} (a_i - a_i^*)\varphi(x_i) \\ b = \frac{1}{n}\sum_{j=1}^{n}\left[y_i - \sum_{i=1}^{n}(a_i - a_i^*)K(x_i, x_j) + \varepsilon\ \text{sgn}(a_j - a_j^*) \right] \end{cases}$$

(6)

Therefore, the line regression Eq. (1) becomes the following explicit form.

$$y = f(x) = \sum_{i=1}^{n}(a_i - a_i^*)K(x, x_i) + b$$

(7)

Based on the nature of the corresponding QP, in general, only a number of coefficients $a_i - a_i^*$ will be assumed as non-zero, and the data points associated with the pair can be referred to as support vectors (SVs).

Fig. 1 shows the topologic structure of SVMs. In Fig. 1, $x = [x_1, x_2, \dots, x_m]^T$.

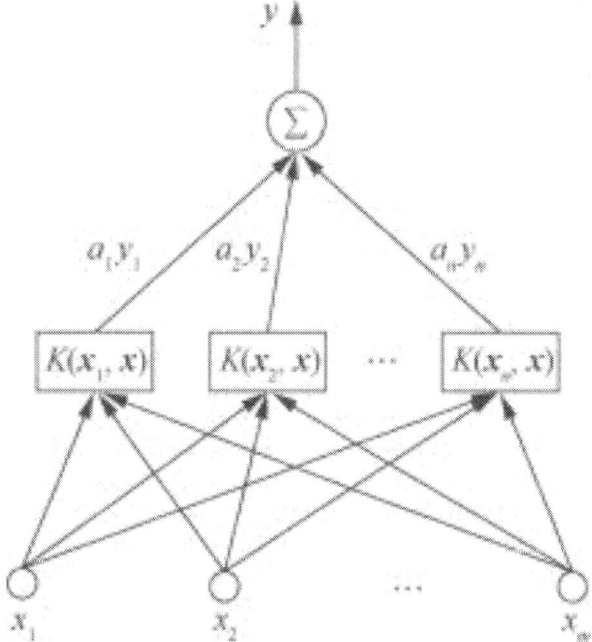

Figure 1: Topologic structure of SVMs.

LS-SVMs

The goal function of SVMs is convex and thus has only one extreme value. However, dimension disaster will arise if the number of training samples is very large, which may result in the optimization algorithm too complex to be carried out. As an extension of the standard SVMs, LS-SVMs are proposed by Suykens and Vandewalle[34] in 1999. The algorithm complexity of LS-SVMs is reduced down greatly by solving linear algebraic equations instead of QP. They have been extensively applied to function approximation and system modeling.

In LS-SVMs, the linear term in the goal function Eq. (4) is replaced by the square term of ξ_i and the inequality constrains Eq. (3) are replaced by equality constrains. Thus, the optimization problem can be written as:

$$\begin{cases} \min_{w,b,\xi,\xi^*} J = \frac{1}{2}\|w\|^2 + \frac{1}{2}c\sum_{i=1}^{n}\xi_i^2 & i = 1,2,\cdots,n \\ \text{s.t.} \quad y_i = w^T\varphi(x_i) + b + \xi_i \end{cases} \tag{8}$$

The Lagrange function is introduced to solve the above equality-constrained optimization problem:

$$L = \frac{1}{2}\|w\|^2 + \frac{1}{2}c\sum_{i=1}^{n}\xi_i^2 - \sum_{i=1}^{n}a_i(w^T\varphi(x_i) + b + \xi_i - y_i) \tag{9}$$

From KKT's condition, one gets the equations:

$$\begin{cases} w = \sum_{i=1}^{n}a_i\varphi(x_i) \\ \sum_{i=1}^{n}a_i = 0 \\ a_i = c\xi_i \quad i = 1,2,\cdots,n \\ w^T\varphi(x_i) + b + \xi_i - y_i = 0 \quad i = 1,2,\cdots,n \end{cases} \tag{10}$$

After eliminating w and ξ_i, the following linear system is obtained.

$$\begin{bmatrix} 0 & \mathbf{1}_n^{\mathrm{T}} \\ \mathbf{1}_n & \Omega + c^{-1}I_n \end{bmatrix} \begin{bmatrix} b \\ a \end{bmatrix} = \begin{bmatrix} 0 \\ y \end{bmatrix}$$

(11)

Where

$$\begin{cases} y = [y_1, y_2, \cdots, y_n]^{\mathrm{T}} \\ \mathbf{1}_n = [1, 1, \cdots, 1]^{\mathrm{T}} \\ a = [a_1, a_2, \cdots, a_n]^{\mathrm{T}} \\ \Omega_{i,j} = K(x_i, x_j) \quad i, j = 1, 2, \cdots, n \end{cases}$$

(12)

Eq. (11) can be solved for the parameters a_i (i = 1,2, ... ,n) and b by use of least squares method. Therefore, the resulting LS-SVM model is given as

$$y = f(x) = \sum_{i=1}^{n} a_i K(x, x_i) + b$$

(13)

As a result of the modifications in LS-SVMs, training requires only to solve the linear equations Eq. (11) instead of the computationally hard quadratic programming problem Eq. (5) in the standard SVMs. However, this is done at expense of losing the sparseness of solution. All the training samples act as support vectors in LS-SVMs.

A Simple Example

In order to validate the learning and generalizing capability of LS-SVMs, a simulation experiment is given as follows.

Consider a two-input one-output nonlinear system:

$$y = \frac{\sin^2 \pi(x_1 - x_2)}{1 + x_1^2 + x_2^2}$$

(14)

It constitutes a spatial curved surface. Taking 200 points on $[-1, 1] \times [-1, 1]$ stochastically, we create training sample data by the following formula.

$$\begin{cases} y = \frac{\sin^2 \pi(x_1 - x_2)}{1 + x_1^2 + x_2^2} + \eta \\ \eta \sim N(0, 0.01^2) \end{cases}$$

(15)

Fig. 2(a) shows the exact curved surface and the training samples.

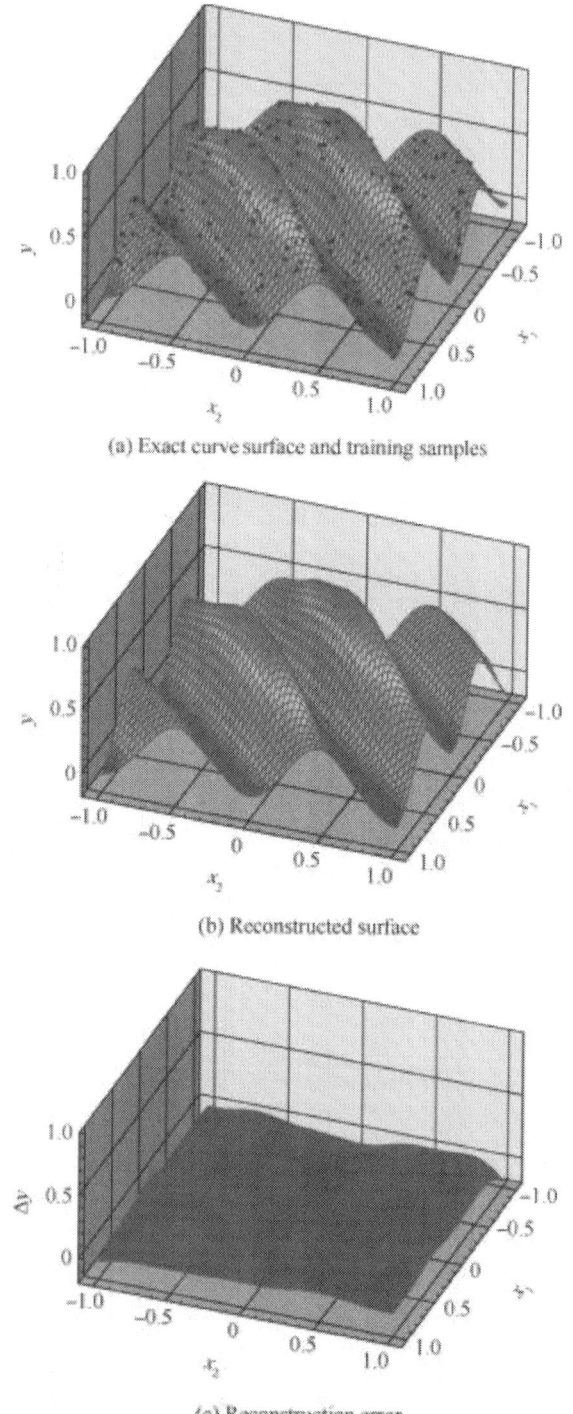

(a) Exact curve surface and training samples

(b) Reconstructed surface

(c) Reconstruction error

Figure 2: LS-SVM modeling results of a two-dimensional nonlinear system.

With the following radius base function adopted, the LS-SVM model is set up for the nonlinear system Eq. (14) from the training samples, where the penalty factor $c = 4$ and the width parameter of radius base kernel function $2\sigma^2 = 0.25$.

$$K(x, x_i) = \exp\left(-\frac{\|x - x_i\|^2}{2\sigma^2}\right)$$

(16)

where σ is kernel width. Then, the surface is reconstructed using the LS-SVM model and shown in Fig. 2(b). The reconstruction error Δy is shown in Fig. 2(c). It is seen that the reconstructed surface approximate the exact one very well. One can expect that the reconstruction error will be reduced down further with decline of the noise level in training samples and/or increase of the number of training samples.

HIGH ANGLE-OF-ATTACK UNSTEADY AERODYNAMIC MODELING METHOD BY USE OF SVMS

As mentioned previously, little understanding of unsteady flow mechanism at high angles of attack leads to great difficulties in model structure determination and parameter estimation when mathematically modeling aerodynamics. The development of SVMs provides a novel tool for unsteady aerodynamic modeling at high angles of attack. Three issues below are involved in high angle-of-attack unsteady aerodynamic modeling by use of SVMs.

Selection of Input Variables
One of the important characteristics of high angle-of-attack aerodynamics is its dependence not only on the instantaneous flight states but also on their time history. Hence, the selection of input variables must enable the SVMs to incorporate the impact of motion history on aerodynamics.

The parameter of reduced frequency k is employed as one of the input variables in most of the previous researches on intelligent modeling such as FL 13· 14 [and] 15 and NNs. [18] For example, Ref. [13] took $(\alpha$, $\dot{\alpha}$, $\ddot{\alpha}$, k, β, $\delta_e)$ as the input variables when modeling aerodynamics of pitching oscillations, while Ref. [14] took $(\alpha, \phi$, $\dot{\phi}, k, \beta, \psi$, $\dot{\psi})$ as the input variables when modeling aerodynamics of yawing and rolling oscillations, where α is angle of attack, β is sideslip angle, ϕ is roll angle, δ_e is elevator deflection angle and ψ is yawing angle. It is noted that the reduced frequency is an unsteadiness parameter adopted specifically in forced-oscillation wind tunnel tests, which does not exist in flight tests. When the aerodynamic models set up from wind tunnel test data are used to predict aerodynamic characteristics in flight tests, an equivalent reduced frequency is needed, which is obtained by fitting a segment of flight test data. For example, the equivalent reduced frequency of pitching motion is calculated by solving an optimization problem [15]:

$$\min_{\alpha_0, \alpha_s, k, \varphi} J = \sum_{i=1}^{n} \{\alpha - [\alpha_0 + \alpha_s \cos(k\tau + \varphi)]\}^2$$

$$+ \sum_{i=1}^{n} \{\bar{\alpha} - [-\alpha_s k \sin(k\tau + \varphi)]\}^2 \tag{17}$$

where $\bar{\alpha}$ is nondimensional rate of angle of attack; τ is nondimensional time; n is the pre-specified number of points for fitting. The mean angle-of-attack α_0, amplitude α_s, reduced frequency k, and phase φ are the parameters to be determined.

A series of problems arise from this processing technique: (a) The nonlinearity of the goal function may lead to local optimality. (b) Singularity may occur in some cases. As a special case, k can be any value at constant angles of attack. (c) For coupled 6-DOF motions, it is not assured whether the equivalent reduced frequency should be determined in three directions of body-axis or in two directions of angle of attack and sideslip angle. (d) The number of points for fitting may have great effects on the resulting equivalent reduced frequency. (e) The resulting equivalent reduced frequency may vary acutely along with time. Fig. 3 shows the time history of the equivalent parameter during a coupled yawing-rolling ramp motion. [18] It skips frequently in the process of ramp (0–3 s) and the reverse (6.7–9.7 s).

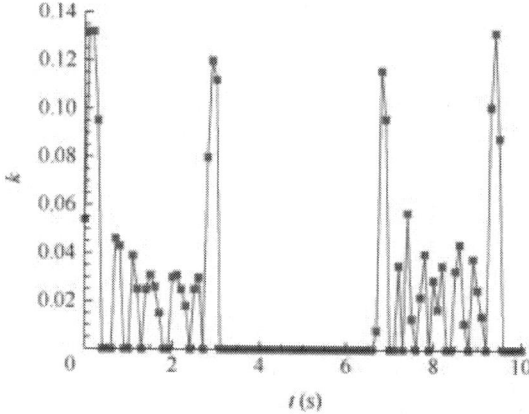

Figure 3: Equivalent reduced frequency of coupled yawing-rolling ramp motion.[18]

As a solution, several sampling points of the current and previous flight states could be employed to describe the effects of motion history. Ref.[16] took $[\alpha(\tau), \alpha(\tau-1), \alpha(\tau-2); \dot{\alpha}(\tau), \dot{\alpha}(\tau-1), \dot{\alpha}(\tau-2); ...]$ as input variables when modeling the longitudinal aerodynamics of a pitching delta wing by use of back propagation neural network (BPNN). Ref.[19] took $[\alpha(\tau), \alpha(\tau-\Delta\tau), \alpha(\tau-2\Delta\tau), \alpha(\tau-3\Delta\tau), \alpha(\tau-4\Delta\tau); \beta(\tau), \beta(\tau-\Delta\tau), \beta(\tau-2\Delta\tau), \beta(\tau-3\Delta\tau), \beta(\tau-4\Delta\tau)]$ as input variables when modeling 6-component aerodynamic coefficients by use of radius base function neural network (RBFNN). The references exhibit satisfying results.

In this way, however, two problems remain to be solved: (a) How long time before do the flight states have no effect upon the current aerodynamics? (b) How to determine the number of sampling points? As for the first question, the time length $m \cdot \Delta\tau$ can be determined according to the state-space models [5 and 6] or the differential-equation models, [7, 8, 9 and 10] to be $(1-2)\, \tau_1$ for instance, where τ_1 is the characteristic time constant in the mathematical models. As for the second question, the number of sampling points m can be determined appropriately according to the above time length and the mode frequencies of the aircraft, to be $[8(m \cdot \Delta\tau) \cdot \max(\bar{\omega}_{SP}, \bar{\omega}_{DR})]$ for instance, where $\bar{\omega}_{SP}$ is the nondimensional frequency of short-period mode and $\bar{\omega}_{DR}$ is that of Dutch roll mode.

Selection of Output Variables

In intelligent modeling of high angle of-attack unsteady aerodynamics, the most direct output variables are aerodynamic force and moment coefficients. There is a fall in this way that the errors in dynamic test data will affect the predictions of static aerodynamics. As we know, dynamic wind tunnel test data include greater errors than static ones in general.

With a view to engineering application, aerodynamics can be decomposed as follows (pitch moment coefficient C_m as an example).

$$C_m = C_{m,st} + \Delta C_{m,dyn} \tag{18}$$

A candidate way is to model the dynamic increment $\Delta C_{m,dyn}$ instead of the aerodynamic coefficient C_m, while the static component $C_{m,st}$ takes the static wind tunnel test results. Thus, all the dynamic test data, measured in different forms of motions and even in different wind tunnels, can be put together for SVM training.

3.3. Determination of SVM parameters

The selection of kernel function and determination of SVM parameters are another important problems for aerodynamic modeling. They also have decisive effects upon the fitting precision, the generalization ability, and the training speed of SVMs.

The kernel function decides the mapping pattern from the input space to the feature space. Theoretically, the kernel can be any symmetry function satisfying Mercer's condition. The commonly used kernels include:

(1) Polynomial function: $K(x, x_i) = (x^T x_i + 1)^q$
(2) Radius base function (RBF): $K(x, x_i) = \exp[-\|x - x_i\|^2/(2\sigma^2)]$
(3) Sigmoid function: $K(x, x_i) = \tanh[r(x^T \cdot x_i) + c]$
(4) Potential function: $K(x, x_i) = \exp[-\|x - x_i\|/(2\sigma^2)]$

There is no universal principle to guide the selection of kernel function. However, the RBF kernel is efficient for nonlinear system modeling, which is validated by a large number of simulation experiments and applications.

Having selected the kernel function (RBF kernel is adopted in unsteady aerodynamic modeling in the next section), the penalty factor c and the kernel width σ in LS-SVMs need also to be determined. The performance of LS-SVMs will be improved by adjusting the two parameters. The methods such as m-fold cross-validation and leave-one-out (LOO) can be utilized to determine these parameters. [35 and 36]

The m-fold cross-validation is the most commonly used method to estimate the generalization error. In this method, the training sample set is first partitioned into m subsets with nearly the same size which do not intersect each other. For given parameters c and σ, one of the subsets is taken as the testing set at a time, the others as the training set. After the parameters a_i ($i = 1,2, \ldots ,n$) and b in the model are obtained by the training algorithm from the training set, the mean square sum of prediction errors $MSSE_i$ can be calculated for the testing set and is taken as the testing error. The average of m testing errors $MSSE = (1/m)\Sigma MSSE_i$ is taken as the generalization error. Thus, the parameters c and σ can be adjusted according to the generalization error.

For a group of parameters c and σ, totally m times of training and testing are needed to calculate the average error MSSE. Consequently, using m-fold cross-validation to optimize the parameters is computationally expensive. The higher m is, the more computation time it costs. One should select an appropriate low value of m to perform m-fold cross-validation, especially in case of large number of training samples.

For unsteady aerodynamic modeling at high angles of attack, it is important that all the data of an individual dynamic test should be taken as a unit and put into same a subset during the m-fold partitioning.

RESULTS AND DISCUSSION

Aerodynamic Modeling of a Pitching Delta Wing

The wind tunnel test data are taken from Ref.[37] The test model is a sharp-edged delta wing with aspect ratio $A = 2$, as shown in Fig. 4. The pitching axis is located at 67%c_0, where c_0 is the wing chord at midspan. The large-amplitude pitching oscillation tests were carried out in the 7 foot × 10 foot (1 foot = 30.48 cm) low-speed wind tunnel at NASA Ames Research Center, with Reynolds number $Re = 4.5 \times 10^5$ and the pitch moment reference point at 77%c_0.

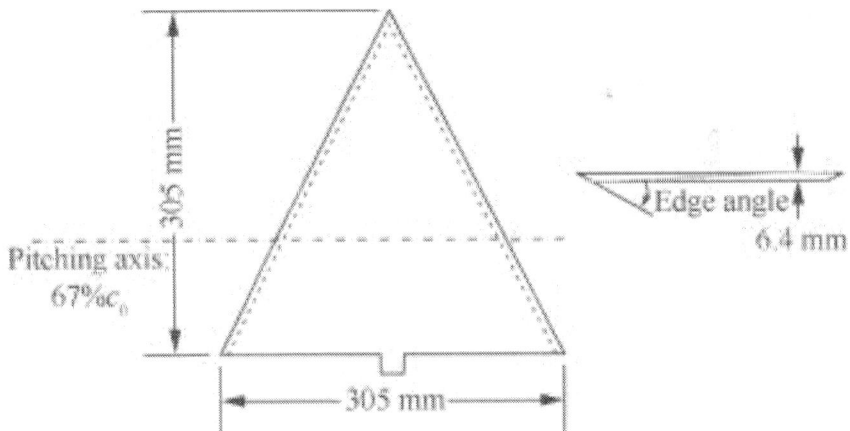

Figure 4: The delta wing model with aspect ratio $A = 2$.

In the forced-oscillations, the time history of angle of attack is

$$\alpha(\tau)=45°-45°\cos(k\tau) \tag{19}$$

where $k = \omega\bar{c}/V$ is reduced frequency, $\bar{c}$ is mean aerodynamic chord, ω is oscillation frequency, V is velocity; $\tau = t(V/\bar{c})$ is nondimensional time.

The nondimensional parameter characterizing unsteadiness $K = 0.01$, 0.02, 0.03, and 0.04, where K is defined as

$$K = \dot{\alpha}_{max} c_0 / (2V)$$

(20)

Converted to the conventional reduced frequency, $k = 0.017$, 0.034, 0.051, 0.068.

The unsteady aerodynamic data are transformed to the dynamic increments:

$$\begin{cases} \Delta C_{L,dyn} = C_L - C_{L,st} \\ \Delta C_{D,dyn} = C_D - C_{D,st} \\ \Delta C_{m,dyn} = C_m - C_{m,st} \end{cases}$$

(21)

At first, the LS-SVM models of $\Delta C_{L,dyn}$, $\Delta C_{D,dyn}$, and $\Delta C_{m,dyn}$ are set up respectively from the dynamic measurement data with $K = 0.01$, 0.03, 0.04, with RBF kernel adopted, and the input variables taken as $\alpha(\tau)$, $\alpha(\tau - 8)$, $\alpha(\tau - 16)$, $\alpha(\tau - 24)$, $\alpha(\tau - 32)$, and $\bar{\bar{q}}(\tau)$, where $\bar{\bar{q}}$ is nondimensional pitching rate. This means that the instantaneous aerodynamics $C_m(\tau)$, for example, depends upon the angle-of-attack history during $[\tau - 32, \tau]$.

The SVM parameters determined by the m-fold cross-validation are listed as follows (with the sample inputs normalized):

$$C_D : \quad c = 3.2, \ 2\sigma^2 = 0.22$$
$$C_L : \quad c = 3.2, \ 2\sigma^2 = 0.21$$
$$C_m : \quad c = 3.2, \ 2\sigma^2 = 0.22$$

Fig. 5 shows the aerodynamic coefficients predicted by the LS-SVM models, in comparison with the measurement results, where the black dashed lines denote the static wind tunnel test data, the colored symbols are the dynamic wind tunnel test data, the colored lines are the predictions of the LS-SVM models, and the arrows indicate the direction of hysteretic loops. The predictions approximate the wind tunnel test data well, which indicates that the LS-SVMs have great learning ability.

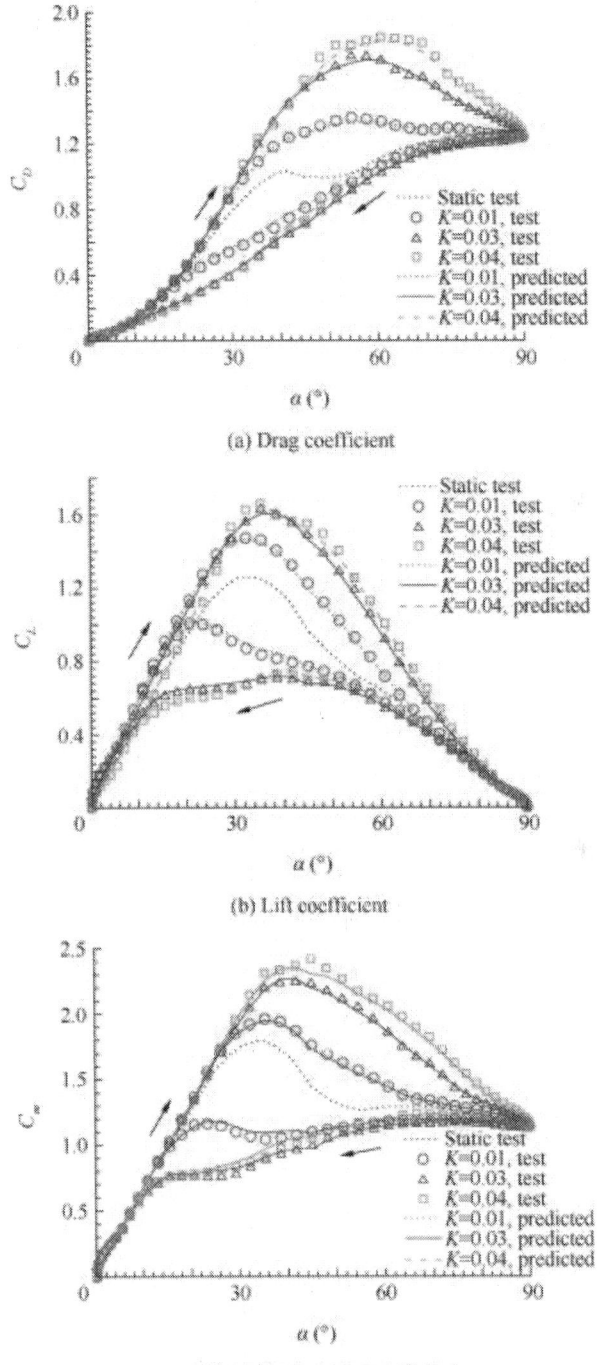

(a) Drag coefficient

(b) Lift coefficient

(c) Pitching moment coefficient

Figure 5: LS-SVM predictions and training data of pitching delta wing.

Subsequently, the LS-SVM models are utilized to predict the aerodynamic characteristics of the large-amplitude pitching oscillation with $K = 0.02$. Fig. 6 shows the predicted aerodynamics in comparison with that obtained by wind tunnel tests. The predictions are in agreement with the test data, in spite of some tolerable discrepancies, which shows that the LS-SVMs have satisfying generalization performance.

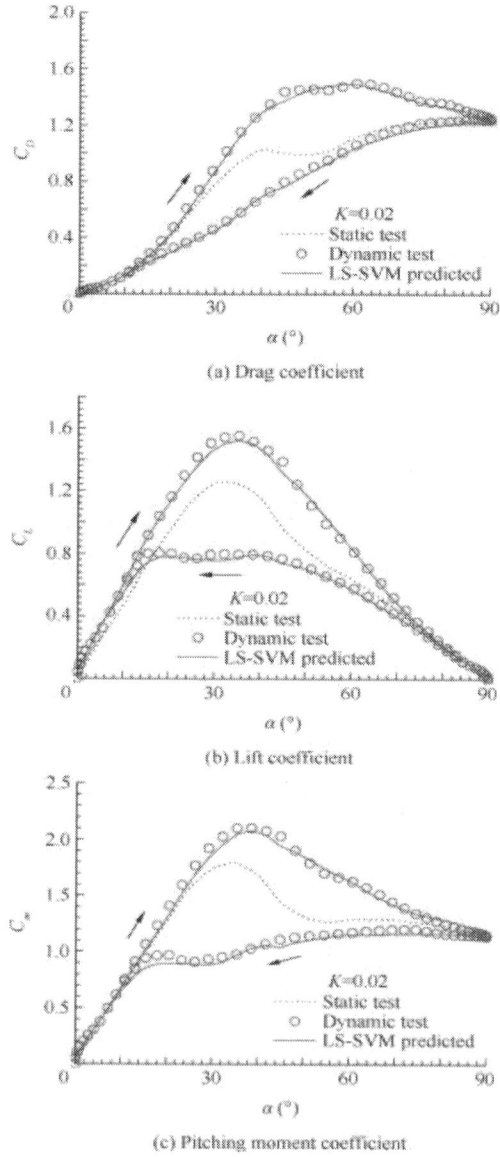

(a) Drag coefficient

(b) Lift coefficient

(c) Pitching moment coefficient

Figure 6: LS-SVM generalization and wind tunnel test data of pitching delta wing.

Aerodynamic Modeling of a Rolling Aircraft Configuration
Ref.[25] presented the rolling oscillation wind tunnel test results of F-16XL.
The tests were executed with a 0.18-scale F-16XL model using a forced-
oscillation rig in the 14 foot × 22 foot subsonic wind tunnel at NASA
Langley Research Center. The test model was forced to roll around the
longitudinal axis at the given pitch angles: $\phi = \phi_s \sin(k\tau\ )$. Here, the
reduced   frequency $k = \omega b'/(2\,V)$,   the   nondimensional   time
$\tau = t(2\,V/b'), b'$ is wing span.

In rolling oscillations, the angle of attack and sideslip angle vary as
follows.

$$\begin{cases} \alpha(\tau) = \tan^{-1}(\tan\theta\cos\phi) \\ \beta(\tau) = \sin^{-1}(\sin\theta\sin\phi) \end{cases} \tag{22}$$

where θ is the pitch angle and ϕ the roll angle.

The roll moment data were obtained at a series of pitching angles, $\theta = 0°$,
10°, 15°, 20°, 25°, 30°, 36°, 40°, 50°, 60°, 70°, and 75°, with the
amplitude $\phi_s = 10°$, 20°, and 30°, and a constant maximum roll rate
of $\bar{P}_{max} = p_{max}b'/(2V) = 0.04$.

The wind tunnel (W.T.) data used here are obtained from digitizing the
AIAA paper. The error of roll moment coefficient C_l in the data points is
less than 0.0005.

Aerodynamic modeling is performed directly for the roll moment
coefficient due to lack of the corresponding static test data. At first, the test
data with $\phi_s = 10°$ and 30° are employed to train the LS-SVM model of C_l,
where RBF kernel is adopted also, and the input variables
take $\alpha\ (\tau), \alpha\ (\tau - 5), \alpha\ (\tau - 10), \alpha\ (\tau - 15), \alpha\ (\tau - 20), \beta\ (\tau), \beta\ (\tau - 5), \beta\ (\tau - 10), \beta\ (\tau - 15), \beta\ (\tau - 20)$, and $\bar{P}\ (\tau)$. The SVM parameters determined
by m-fold cross-validation are (with the sample inputs normalized): $c = 5.0$
and $2\sigma^2 = 0.58$. The roll moment coefficient predicted by the LS-SVM
model, as well as the one obtained from wind tunnel measurement, is
presented in Fig. 7, where the red circles are the dynamic wind tunnel test
data, the blue solid lines denote the predictions of the LS-SVM model, and
the arrows indicate the direction of hysteretic loops. The figure shows a
satisfactory approximation.

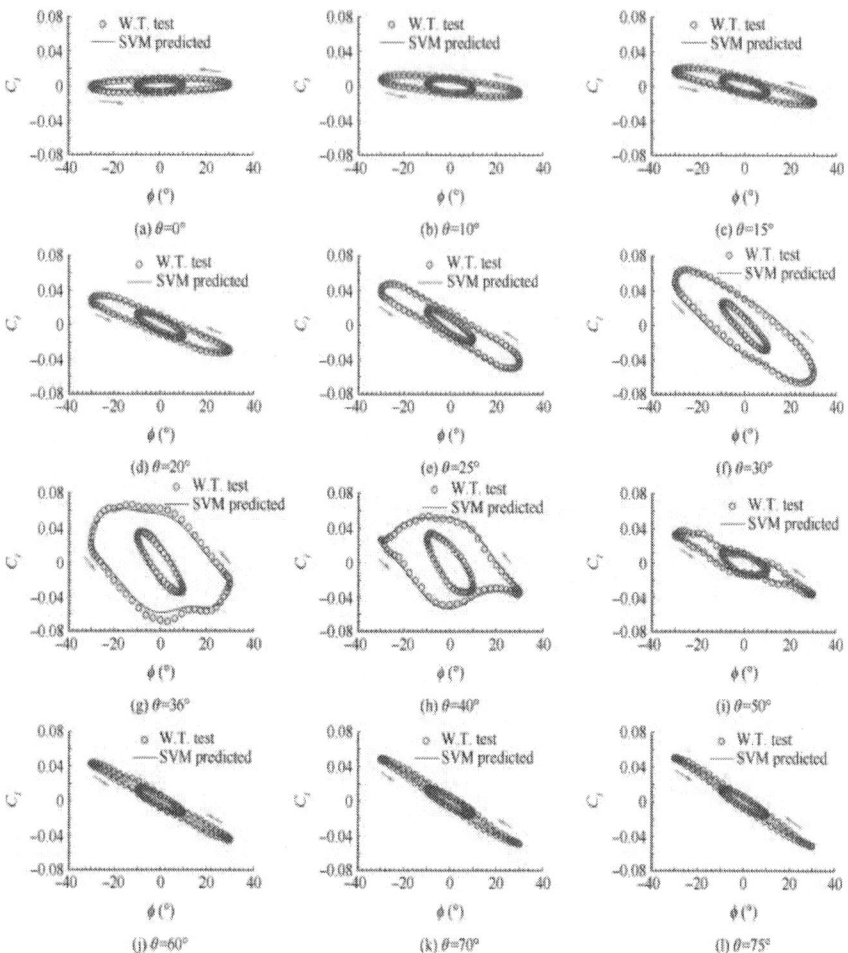

Figure 7: Prediction results of LS-SVM model and training data for rolling aircraft configuration.

The LS-SVM model is then employed to predict the aerodynamic response to the rolling oscillations with the amplitude $\phi_s = 20°$. Fig. 8 shows the predictions in comparison with the wind tunnel test data. It can be seen that there are some tolerable discrepancies in the range of $\alpha_0 = 30°–50°$, where the unsteady aerodynamic effects are extremely great. Nevertheless, the overall agreement indicates the satisfying generalization performance of LS-SVMs.

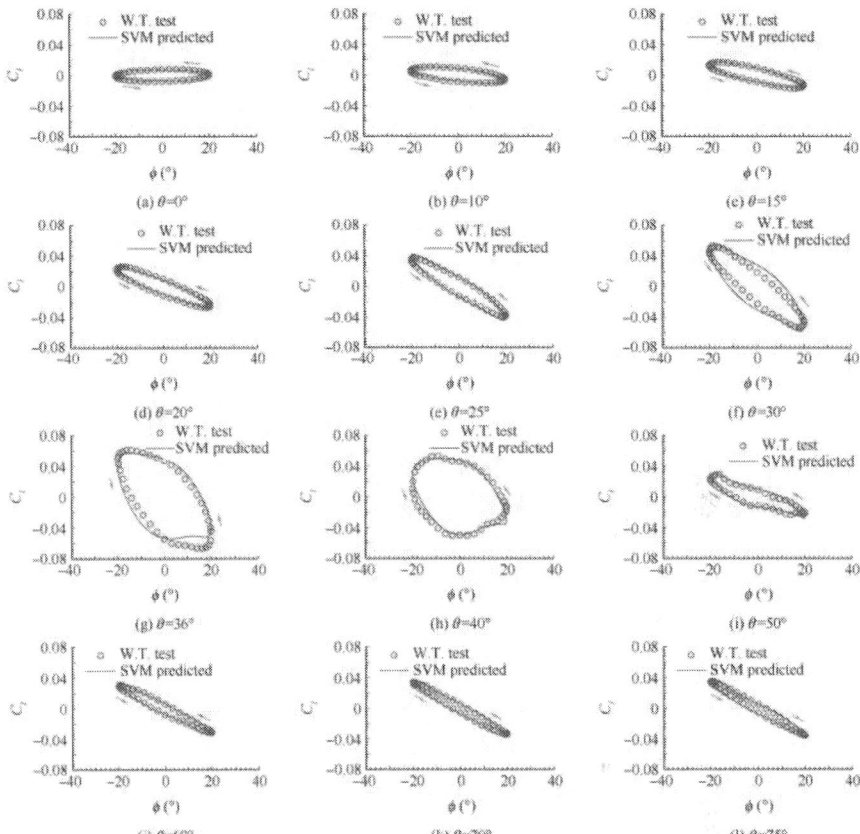

Figure 8: Generalization results of LS-SVM model and wind tunnel data for rolling aircraft configuration.

CONCLUSIONS

SVMs are a novel type of machine learning method developed on the basis of statistical learning theory, embodying the structural risk minimization principle. They are gaining popularity due to the features such as simple structure, global optimality and empirical performance. The following can be concluded from the presentations and applications in this paper.

(1) The aerodynamic modeling results of the pitching delta wing and the rolling aircraft configuration show that the LS-SVMs have excellent learning capability and satisfying generalization performance, and thus become an attractive means in the field of high angle-of-attack unsteady aerodynamic modeling.

(2) In order to describe the effects of motion history on aerodynamics, one can take several sampling points of the current and previous flight states as the inputs of SVMs. It is suggested to model the dynamic increments instead of aerodynamic coefficients, for the static wind tunnel test data have higher precision than dynamic ones in general.

(3) RBF kernel is an appropriate selection for unsteady aerodynamic modeling at high angles of attack. The penalty factor c and the kernel width σ can be determined by means of m-fold cross-validation. It is important that all the data of an individual dynamic test should be taken as a unit and put into same a subset during the m-fold partitioning.

ACKNOWLEDGMENTS

The authors are grateful to all the members of their work group for help in the present research. They also thank the anonymous reviewers for their constructive review of the manuscript. This study was supported by a Chinese Government Contract.

REFERENCES

1. Lin GF, Lan CE. A generalized dynamic aerodynamic coefficient model for flight dynamics applications. Reston: AIAA; 1997. Report No.: AIAA-1997-3643.

2. Tobak M, Schiff LB. Aerodynamic mathematical modeling-basic concepts. Moffett Field, CA:NASA Ames Research Center; 1981. Report No.: 19810022564.

3. Chin S, Lan CE. Fourier functional analysis for unsteady aerodynamic modeling. AIAA J 1992;30(9):2259–66.

4. Murphy PC, Klein V. Estimation of aircraft nonlinear unsteady parameters from dynamic wind tunnel testing. Reston: AIAA; 2001. Report No.: AIAA-2001-4016.

5. Goman M, Khrabrov A. State-space representation of aerodynamic characteristics of an aircraft at high angles of attack. J Aircraft 1994;31(5):1109–15.

6. Lutze FH, Fan YG, Stagg G. Multi-axis unsteady aerodynamic characteristics of an aircraft. Reston: AIAA; 1999. Report No.: AIAA-1999-4011.

7. Abramov NB, Goman MG, Khrabrov AN. Aircraft dynamics at high incidence flight with account of unsteady aerodynamic effects. Reston: AIAA; 2004. Report No.: AIAA-2004-5274.

8. Abramov N, Goman M, Demenkov M, Khrabrov A. Lateraldirectional aircraft dynamics at high incidence flight with account of unsteady aerodynamic effects. Reston: AIAA; 2005. Report No.: AIAA-2005-6331.

9. Wang Q, Cai JS. Unsteady aerodynamic modeling and identification of airplane at high angles of attack. Acta Aeronaut Astronaut Sin 1996;17(4):391–8 Chinese.

10. Wang Q, He KF, Qian W, Mao ZJ. Aerodynamic modeling of spatial maneuvering aircraft at high angles of attack. Acta Aeronaut Astronaut Sin 2004;25(5):447–51 Chinese.

11. Huang XZ. Nonlinear indicial response/internal state-space representation and its application on delta wing aerodynamics. Reston: AIAA; 2003. Report No.: AIAA-2003-3944.

12. Pashilkar AA. Flight dynamic analysis of the flow incidence rate model. Reston: AIAA; 2002. Report No.: AIAA-2002-0098.

13. Wang Z, Lan CE, Brandon JM. Fuzzy logic modeling of nonlinear unsteady aerodynamics. Reston: AIAA; 1998. Report No.: AIAA- 1998-4351.

14. Wang Z, Lan CE, Brandon JM. Fuzzy logic modeling of lateraldirectional unsteady aerodynamics. Reston: AIAA; 1999. Report No.: AIAA-1999-4012.

15. Wang Z, Lan CE. Unsteady aerodynamic effects on the flight characteristics of an F-16XL configuration. Reston: AIAA; 2000. Report No.: AIAA-2000-3901.

16. Rokhsaz K, Steck JE. Application of artificial neural networks in nonlinear aerodynamics and aircraft design. J Aerospace 1993;102:1790–8.

17. Soltani MR, Sadati N, Davari AR. Neural network: a new prediction tool for estimating the aerodynamic behavior of a pitching delta wing. Reston: AIAA; 2003. Report No.: AIAA- 2003-3793.

18. Shi ZW, Wang ZH, Li JC. The research of RBFNN in modeling of nonlinear unsteady aerodynamics. Acta Aerodyn Sin 2012;30(1): 108–12 Chinese.

19. Wang Q, He KF, Qian WQ, Zhang TJ, Cheng YQ. Unsteady aerodynamics modeling for flight dynamics application. Acta Mech Sin 2012;28(1):14–23.

20. Wang Q, Wu KY, Zhang TJ, Kong YN, Qian WQ. Aerodynamic modeling and parameter estimation from the QAR data of an airplane approaching a high-altitude airport. Chin J Aeronaut 2012;25(3):361–71.

21. Singh R, Baeder J. Direct calculation of three-dimensional indicial lift response using computational fluid dynamics. J Aircraft 1997;34(4):465–71.

22. Ghoreyshi M, Cummings RM, DaRonch A, Badcock KJ. Transonic aerodynamic load modeling of X-31 aircraft pitching motions. AIAA J 2013;51(10):2447–64.

23. Ghoreyshi M, Jirasek A, Cummings RM. Reduced order unsteady aerodynamic modeling for stability and control analysis using computational fluid dynamics. Prog Aerosp Sci 2014;71:167–217.

24. Cooperative programme on dynamic wind tunnel experiment for manoeuvering aircraft. Reston: AIAA; 1996. Report No.: AGARD-AR-305.

25. Brandon JM, Foster JV. Recent dynamic measurements and considerations for aerodynamic modeling of fighter airplane configurations. Reston: AIAA; 1998. Report No.: AIAA-1998- 4447.

26. Huang D. Unsteady aerodynamic characteristics for the aircraft oscillation in large amplitude [dissertation]. Nanjing: Nanjing University of Aeronautics and Astronautics; 2007 [Chinese].

27. 27.Findlay D, Guruswamy G. Numerical analysis of aircraft high angle of attack unsteady flows. Reston: AIAA; 2000. Report No.: AIAA-2000-1946.

28. Schu¨tte A, Einarsson G, Raichle A, Schoning B, Mo¨nnich W, Forkert T. Numerical simulation of manoeuvreing aircraft by aerodynamic, flight mechanics, and structural mechanics coupling. J Aircraft 2009;46(1):53–64.

29. Paul R, Murua J, Gopalarathnam A. Unsteady and post-stall aerodynamic modeling for flight dynamics simulation. Reston: AIAA; 2014. Report No.: AIAA-2014-0729.

30. Gunn SR. Support vector machines for classification and regression. Southampton (UK): University of Southampton; 1998. Report No.: ISIS-1-98.

31. Scho¨lkopf B, Sung K-K, Burges CJC, Girosi F, Niyogi P, Poggio T, et al. Comparing support vector machines with Gaussian kernels to radial basis function classifiers. IEEE Trans Signal Process 1997;45(11):2758–65.

32. Fan H-Y, Dulikravich GS, Han Z-X. Aerodynamic data modeling using support vector machines. Reston: AIAA; 2004. Report No.: AIAA-2004-0280.

33. Vapnik V. The nature of statistical learning theory. New York: Springer-Verlag; 1995.

34. Suykens JAK, Vandewalle J. Least squares support vector machine classifiers. Neural Processing Letters 1999;9(3):293–300.

35. Chapelle O, Vapnik V, Bousqet O, Mukherjee S. Choosing multiple parameters support vector machines. Machine Learning 2002;46(1):131–60.

36. Deng NY, Tian YJ. A new method for data mining: support vector machines. Beijing: Science Press; 2004, p. 355-66 Chinese.

37. Jarrah MA, Ashley H. Impact of unsteadiness on maneuvers and loads of agile aircraft. Reston: AIAA; 1989. Report No.: AIAA- 1989-1282-CP.

CITATION

Qing Wang, Weiqi Qian, Kaifeng He, Unsteady aerodynamic modeling at high angles of attack using support vector machines, Chinese Journal of Aeronautics, Volume 28, Issue 3, June 2015, Pages 659-668, ISSN 1000-9361, http://dx.doi.org/10.1016/j.cja.2015.03.010.

CHAPTER 9

A Spongy Icing Model for Aircraft Icing

Li Xin[a] , Bai Junqiang[a] , Hua Jun[b] , Wang Kun [a] , Zhang Yang [a]

[a] School of Aerodynamics, Northwestern Polytechnical University, Xi'an 710072, China
[b] Chinese Aeronautical Establishment, Beijing 100012, China

ABSTRACT

Researches have indicated that impinging droplets can be entrapped as liquid in the ice matrix and the temperature of accreting ice surface is below the freezing point. When liquid entrapment by ice matrix happens, this kind of ice is called spongy ice. A new spongy icing model for the ice accretion problem on airfoil or aircraft has been developed to account for entrapped liquid within accreted ice and to improve the determination of the surface temperature when entering clouds with supercooled droplets. Different with conventional icing model, this model identifies icing conditions in four regimes: rime, spongy without water film, spongy with water film and glaze. By using the Eulerian method based on two-phase flow theory, the impinging droplet flow was investigated numerically. The accuracy of the Eulerian method for computing the water collection efficiency was assessed, and icing shapes and surface temperature distributions predicted with this spongy icing model agree with experimental results well.

INTRODUCTION

Ice accretion is a common and an important feature in flight. The presence of ice accretion can cause a serious safety problem; the most severe penalties encountered deal with increased drag, decreased lift, decreased stall angle, and reduced controllability. On the safety issues, aircraft icing has caused more than 50 accidents in the US in recent 20 years. The ice accretion problems can be studied by the wind tunnel testing and the real flight test, or the engineering and numerical approaches. The real flight test and the wind tunnel testing require extensive analyses, which are expensive and dangerous. The engineering approach employs empirical formulation and experimental data, which is much simpler but lack of precision. Therefore, the numerical method is widely adopted for its economical, efficient, and accurate features.

Conventionally, the simulation of ice accretion is based on the Lagrangian particle-tracking technique for the trajectory calculation and employing Messinger icing model. Bougault et al. developed an Eulerian method for the ice accretion.[1] After that, Eulerian method became popular because of its simplicity and efficiency. However, better numerical methods are needed to track and collect droplets and further studies are needed to understand the key factors on the icing growth. Currently, a number of ice accretion codes have been well developed by some international icing communities, such as ONERA (France), CIRAML (Italy), DRA (United Kingdom), LEWICE (USA), and FENSAP-ICE (Canada). The FENSAP-ICE code employs the Eulerian method, and the others are using the Lagrangian method.

The ice accretion problem has been studied widely in the last several decades, and many work had been done on prediction of ice shape by many researchers such as Bragg, Shen, and Shin and Bond.[2, 3 and 4] In 1985, Bragg made some improvements on his previous model,[5] derived a method to solve the droplet trajectories and gave some recommendations for further improvement of the method. LEWICE was developed by the University of Dayton in 1983. Potapczuk and Bidwell extended the original LEWICE based on potential flow analysis to LEWICE/NS which used the solution of 2D Navier–Stokes equations.[6] Besides, it has been implemented to solve the energy equation to obtain the heat-transfer coefficient automatically. Another icing code from DRA has also been developed, especially for the ice protection on airfoils and on rotor blades of a helicopter. Recently, an ALE mesh movement scheme for long-term in-flight ice accretion has been performed by Fossati et al.[7]

The Lagrangian approach provides a good result but there are still some shortages. For example, it is hard to be applied on some complicated geometries, such as 3-D aircraft wing and multi-element airfoils. Bourgault developed an Eulerian approach on ice accretion[1] to overcome the shortages. On modeling the aircraft icing, the classical one was developed by Messinger.[8] Then Myers made some improvements on Messinger model.[9 and 10] Also, Bourgault and Otta et al. developed icing model for aircraft respectively.[11 and 12] With regard to the existing codes, FENSAP-ICE employs the icing model developed by Bourgault, ICECREMO employs Myers's lubrication type model, and the icing model of LEWICE is based on Messinger's work.

The icing model developed in this paper included two physical phenomena that traditional icing model ignores: one is the supercooling of accreting ice surfaces, and the other is the sponginess of ice. Experiments have already confirmed that the accreting surface temperature would fall below the freezing point when ice started to form.[13] Karev et al. investigated the icing growth on a cylinder by the wind tunnel testing,[14] and the surface temperature was found below the freezing point of water. The possibility of liquid entrapment by an ice accretion's growing matrix has recognized by List.[13] When droplets were entrapped in the ice matrix, this type of icing is named as the spongy icing. To account for the sponginess and the supercooling, Blackmore and Lozowski[15] proposed a theoretical spongy spray icing model with surficial structure in the field of atmospheric icing. The supercooling of the water film and sponginess of ice affect the ice accretion process, and determine the ice growth rate and the ice shape, so the two physical phenomena which are usually ignored by traditional icing models must be considered in aircraft icing. Besides, like many other icing model, the spongy icing model is based on continuity equation and heat balance equation, so it has good succession.

DESCRIPTION OF THE PROBLEM

To verify and validate the Eulerian droplet tracking method and the collection efficiency, a suitable test case must be considered. A comparison between the Eulerian method and LEWICE is presented. In addition, a comparison with experimental impingement data from Papadakis et al.[16] is discussed as well. The impingement data is presented in the form of LWC distribution and water collection efficiency. The experiments were performed with different MVDs for a NACA23012 airfoil and an iced NACA23012 airfoil. The test conditions are shown in Table 1. For the NACA23012 airfoil, the selected MVDs are 20 and 111 μm, while for the iced NACA23012 airfoil, the MVDs are 20, 52, and 111 μm. Moreover, the Eulerian approach and the icing model were validated by simulating icing conditions on a NACA0012 airfoil. The test temperatures are −4.4, −10, −13.3, and −19.4 °C, which cover both dry and wet regimes; details of the test conditions are given in Table 2. The predicted ice shapes were compared to the experimental results under the same conditions in the NASA Lewis Icing Research Tunnel and the numerical results from LEWICE.[4] At last, the prediction accuracy of the surface temperature was evaluated through the simulation on the surface of a non-rotating horizontal cylinder, which is 0.038 m in diameter. The computational results were compared with the experimental results in Ref.[14]; the test conditions for cylinder are shown in Table 3. Two duration times of ice accretion are set 15 and 15.4 min. The 15 min computation is used to compare ice shape with experiment, while 15.4 min computation is used to validate the surface temperature with experiment. In tables, AOA,V_∞, LWC, p_∞ and MVD are angle of attack, free-stream velocity, liquid water content, free-stream pressure and mean volumetric diameter respectively.

Table 1: Test conditions for droplets impingement.

Parameter	Value
AOA (°)	2.5
V_∞ (m/s)	78.22
LWC (g/m³)	0.5
Chord (m)	0.9144
p_∞ (kPa)	94.8

Table 2: Test conditions for ice accretion on NACA0012.

Parameter	Value
Accretion time (s)	360
AOA (°)	4
MVD (μm)	20
V_∞ (m/s)	67.05
LWC (g/m³)	1
Chord (m)	0.5334
p_∞ (kPa)	101.3

Table 3: Test conditions for ice accretion on cylinder.

Parameter	Value
T (°C)	17
MVD (μm)	43.4
V_∞ (m/s)	30
LWC (g/m³)	3
p_∞ (kPa)	101.3

FORMULATION AND NUMERICAL METHOD

Based on the multiphase flow theory,[17] an Eulerian method to numerically simulate ice accretions is presented in this paper. Some assumptions are established as follows[1]:

(1) Droplets are spherical and complete without any breakage or deformation.
(2) No droplet collision/coalescence.
(3) No heat and mass transfer between the air phase and droplet phase.
(4) No turbulence effect on droplets.
(5) Gravity, air drag, and buoyancy are considered.

Governing Equations for Air Phase

The flow field for air can be obtained by solving the Reynolds-Averaged Navier–Stokes equations. Numerical approach is based on the finite volume form of the integral equations. In a domain of a volume Ω with boundary Ω_a, let ρ, u, v, E, H and p be the density, Cartesian velocity components, total energy, total enthalpy, and pressure, the equation can be written in the integral form:

$$\frac{\partial}{\partial t} + \int\int\int_{\Omega} Q dV + \int_{\partial\Omega} F_c \cdot n dS = \frac{1}{R_e} \int_{\partial\Omega} F_v \cdot n dS \tag{1}$$

The vector of the conserved variables, the convective flux term, and the viscous flux term are given as follows

$$Q = \begin{bmatrix} \rho \\ \rho u \\ \rho v \\ \rho E \end{bmatrix}$$

$$F_c = \begin{bmatrix} \rho u & \rho v \\ \rho u^2 + P & \rho u v \\ \rho u v & \rho v^2 + P \\ \rho u H & \rho v H \end{bmatrix}$$

$$F_v = \begin{bmatrix} 0 & 0 \\ \tau_{xx} & \tau_{xy} \\ \tau_{yx} & \tau_{yy} \\ \Pi_x & \Pi_y \end{bmatrix}$$

$$\text{with} \begin{cases} \Pi_x = u\tau_{xx} + v\tau_{xy} - q_x \\ \Pi_y = u\tau_{yx} + v\tau_{yy} - q_y \end{cases}$$

where τ_{xx}, τ_{xy}, τ_{yx}, τ_{yy} are the elements of the shear-stress tensor, q_x and q_y the elements of heat-flux vector.

The cell-centered finite volume method is employed to solve the Navier–Stokes. The lower–upper symmetric Gauss–Siedel (LU-SGS) algorithm for time marching and Roe scheme for the spatial discretization of the convective flux were implemented in codes. The treatment of the far field boundary condition is based on the introduction of Riemann invariants for a one-dimensional flow normal to the boundary. Isothermal wall boundary condition is applied. Chi et al.[18] suggested that Spalart–Allmaras was proper for simulating ice accretion in the air. The Navier–Stokes equations were closed by the S–A model with thermally perfect gas and temperature-dependent properties.

Governing Equations for Droplets

The droplets distributed in the flow field can be regarded as a kind of pseudo fluid which penetrates in the air flow field. The governing equations for droplets can be written as follows

$$\frac{\partial \rho_d \varphi}{\partial t} + \mathrm{div}(\rho_d V_d \varphi) = S_\varphi \tag{2}$$

where $\varphi = \begin{bmatrix} 1 & v_{dx} & v_{dy} \end{bmatrix}$ denoting the variables in the droplet continuity equation and momentum equation respectively, ρ_d the droplet apparent density (the mass of droplets per unit volume) and $\rho_d = a_v \rho_w$, where a_v and ρ_w denote the droplet volume fraction and the density of water respectively, V_d denotes the droplet velocity vector and S_φ the source term. The water collection efficiency on the surface of the leading edge is solved by the velocity distribution in the air flow field of the airfoil. The finite volume method is applied to discretizing the governing equations, the convective term is discretized using the quadratic upwind interpolation of convective kinematics (QUICK) scheme, and the deferred correction method is used to ensure the diagonal dominance in the discretized equations. The temporal term is discretized using the implicit scheme. The alternating direction implicit (ADI) iteration method is utilized to solve the algebraic equations. In order to guarantee the stability of the iterative solution of algebraic equations, the source term is disposed linearly.

A permeable wall boundary condition is applied to simulating the droplets impingement onto the wall. The far field boundary conditions for the flow field are introduced as follows: $\rho_d = \mathrm{LWC}$, $v_{dx} = u_\infty$, $v_{dy} = 0$, where LWC represents the liquid water content.

On meshing, a single-zone C-type grid with wake cut is developed for the NACA23012 airfoil, the iced NACA23012 airfoil, and the NACA0012 airfoil; the grid has 601×177 grid points (see Figs. 1(a)–(c)) and extends 20 chord lengths away from the airfoil in each direction. A distance to the first grid point off the airfoil surface of 4×10^{-6} m is used to give a y^{+} value of 0.5 approximately. A single-zone O-type grid is developed for cylinder; the grid has 601×165 grid points (see Fig. 1(d)) and extends 20 chord lengths away from the wall in all directions. A distance to the first grid point off the surface of 4×10^{-6} m is used to obtain the y^{+} value between 0.1 and 0.8.

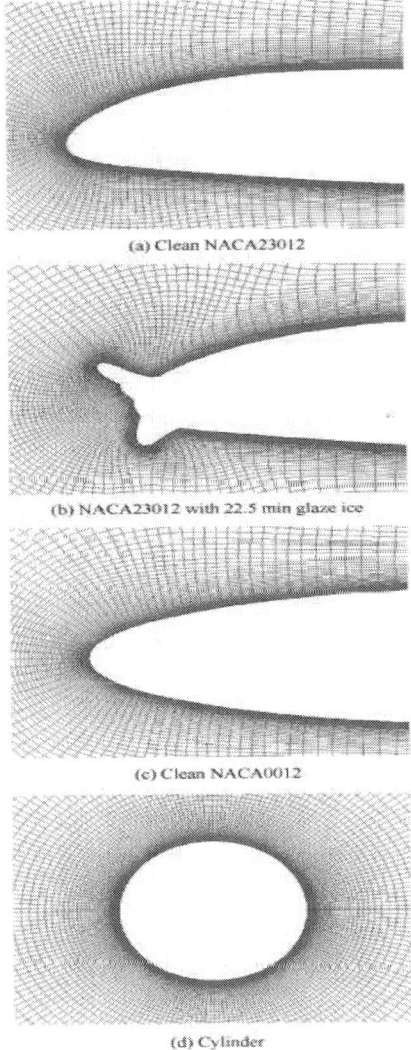

(a) Clean NACA23012

(b) NACA23012 with 22.5 min glaze ice

(c) Clean NACA0012

(d) Cylinder

Figure 1: Grid used in the present study.

CONCEPT AND STRUCTURE OF ICING MODEL

The following assumptions are considered; further experimental verification is needed to validate these assumptions.

(1) Each dendrite has a hemispherical tip and a cylindrical body.
(2) Ice accretes like dendrite crystals from the surface[19] (see Fig. 2).

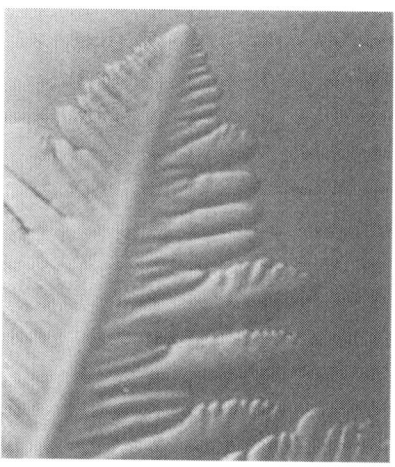

Figure2: Microcosmic photograph of dendrite crystal.

(3) The radius of the hemispherical tip of the dendrite is proportional to the height of ice layer.[20]
(4) The latent heat of solidification produced by the ice forming layer is rejected to the water film layer.
(5) The radius of the dendrite is determined by the ambient temperature and the LWC.[19]However, when the LWC is low, the ambient temperature dominates.

Modeling of Laminar Layer Film

The continuity equation for each control volume can be expressed as follows:

$$R_{in} + R_{imp} = R_{wf\text{-}ifl} + R_{out} \tag{3}$$

where R_{in} is the total mass flux entering the control volume from the previous control volume, R_{imp} the flux of impinging droplets, and R_{out} the total mass flux flow out of the control volume. The mass flux of water transferred from the laminar layer to ice forming zone $R_{wf\text{-}ifl}$, equals the ice accretion flux $R_{wf\text{-}ifl} = I_0 + R_0$. The mass flux entrapped by the ice matrix from the ice forming zone $R_{ifl\text{-}si}$ equals $R_{wf\text{-}ifl}$.

Dukler and Bergelin used the universal velocity distribution equation for turbulent flow of Von Karman in which the non-dimensional film flow rate was expressed by[21]:

$$u^+ = u_f / u_*$$
(4)

where u_f is the velocity of water films, and u_* the friction velocity. The non-dimensional coordinate normal to the substrate is

$$\zeta = \frac{\sqrt{y_t^3 g \rho_w}}{\mu_w}$$
(5)

where y_t is the water film thickness.

The Prandtl mixing length hypothesis results in several equations that are used to describe the velocity profile with the non-dimensional variable given in Eq. (5). The laminar layer is expressed by

$$u^+ = \zeta \quad (0 < \zeta \leq 5)$$
(6)

The total mass flux in the film can be written as

$$R_t = \mu_w f(\zeta) L_{en}^{-1}$$
(7)

where L_{en} is the length of the control volume, R_t the total mass flux in the film

$$f(\zeta) = 0.5\zeta^2 \quad (0 \leq \zeta \leq 5)$$
(8)

For aircraft icing, $\zeta < 5$, which means the water film is laminar. In order to predict the pure ice growth rate and the liquid entrapment within the ice matrix, more details are needed to examine the ice forming layer. The assumptions in this section are applied. Further experimental verifications are needed to validate these assumptions.

Modeling of Ice Forming Layer

Dendritic growth at the ice surface leads to a spongy ice. In List's icing model for a growing hailstone, there is a layer of ice lying between the spongy ice matrix and the surficial liquid. Here, we name this as an ice forming layer, which is assumed to contain some ice crystals that are growing with a uniform tip growth rate. Ice forming layer consists of water and dendritic ice, the thickness of the water film is $y_2 - y_1$ shown in Fig. 3, and all the temperatures are in degrees Celsius.

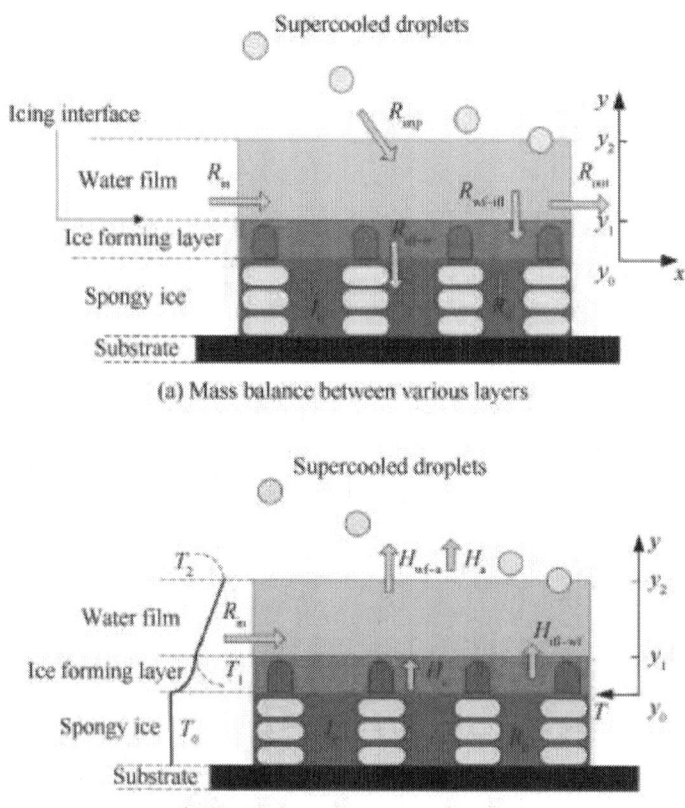

(a) Mass balance between various layers

(b) Heat balance between various layers

Figure. 3: Schematic of the model's surficial structure.

The conduction and evolution of heat of the ice forming zone are described by a 1-D steady-state diffusion equation, known as

$$\frac{d}{dy}\left(k\frac{dT(y)}{dy}\right) + q_v = 0 \tag{9}$$

where y is the coordinate normal to the surface with origin $y_0 = 0$ at the interface, $T(y)$ the temperature in the ice forming layer, k the thermal conductivity of the ice forming layer, which is assumed to be independent of y, with the boundary condition $T = 0$ °C and $dT/dy = 0$ when $y = 0$. The solution for Eq. (9) is

$$T_0 - T_1 = q_v(y_1 - y_0)^2/2k \qquad (10)$$

The volumetric rate of evolution of latent heat in the ice forming layer can be written as

$$q_v = \frac{q_1}{y_1 - y_0} \qquad (11)$$

where q_1 is the latent heat from the ice forming layer to the laminar layer, which can be expressed by

$$q_1 = I_0 L_f - c_w(T_0 - T_1)(I_0 + R_0) \qquad (12)$$

where c_w is the specific heat capacity of pure water at 0 °C(4.2×10^3 J · kg^{-1} · K^{-1}), and L_f is the specific latent heat of fusion of pure water at 0 °C(3.34×10^5 J · kg^{-1}).

The crystal growth theory pointed out that the growth rate at the tips of dendrites is a function of the temperature difference of the bulk liquid into which the dendrite is growing[19]; the growth rate can be expressed as

$$V_c = a\Delta T^b \qquad (13)$$

where a and b are empirical coefficients and ΔT is the temperature difference, $\Delta T = T_0 - T_1$. Like List, [13] we assume that the growth rate of the icing surface is a function of the temperature drop across the ice forming layer, Tirmizi recommended values a and b of 1.87×10^{-4} and 2.09. However, in order to consider the influence of the ambient temperature and to apply this model to aircraft icing field successfully, after calibration work, we recommend b as of $-3.79 \times 10^{-4}T_a^3 - 1.843 \times 10^{-2}T_a^2 - 0.15406T_a + 1.6569$, where T_a is the temperature of air flow and b a sensitive function of the temperature, which has a decisive effect on the speed of the icing interface, and also plays an important role in dividing the type of ice accretion.

Assume V_c to be the rate of advance of liquid across the icing interface, which means

$$V_c = V_1 = \frac{I_0 + R_0}{\rho_w} \qquad (14)$$

To determine the thickness of the ice forming layer is very important for modeling the ice forming layer. Tirmizi assumed that there is a continuous linear relationship between temperature drop and the radius of curvature of the tip of a growing ice dendrite, such as

$$r_c = c + d/\Delta T \tag{15}$$

Tirmizi recommended applying $c = 6.16 \times 10^{-5}$ and $d = 2.024 \times 10^{-5}$ for dendritic ice, but for aircraft icing we recommended applying $c = 4.36 \times 10^{-5}$ and $d = 2.864 \times 10^{-5}$. The main function of the two parameters is to make sure that the iterative solution of the spongy icing model is convergent.

The thickness of ice forming layer is proportional to the radius of freely-growing ice dendrite tips

$$y_1 - y_0 = k_r r_c \tag{16}$$

where k_r is the factor of proportionality; for aircraft icing, recommend $k_r = 1.32$. The main function of k_r is not only to guarantee iterative solution of the spongy icing model is convergent but also making sure the predicted ice shapes have physical significance. K_r is dependent of velocity and liquid water content. In recent years, a wide variety of tests have been performed in the NASA Lewis icing research tunnel (IRT). [22] These data facilitates a systematic calibration of the parameters mentioned above, and the parameters are calibrated by experimental data of different temperatures, liquid water contents, MVD, and different speeds. After the calibration work, c and d are constant, no matter for the rime, the glaze or the spongy regime, but b depends on ambient temperature, for b is a function of temperature. From Eqs.(15) and (16) we can conclude that fine ice crystals are (grow at high film supercoolings) with tips of smaller radius of curvature than coarser ice crystals (grow at lower supercoolings). Experiments have proved that high temperature (low film supercoolings) produced coarse ice columnar crystals and vice versa. [20]

Heat Balance of Laminar Layer Film
The heat balance for the laminar layer is

$$H_{in} + H_{ifl-wf} + H_{wf-a} + H_{out} = 0 \tag{17}$$

where H_{in} is the sensible heat flux in the laminar layer which is used to heat the inflowing liquid, and H_{ifl-wf} the bulk heat flux from the ice forming layer to the laminar layer, the initial value of H_{out} is zero, and the value will be updated during iteration process. H_{wf-a} is the heat flux from the laminar layer to the air flow.

The expression for the H_{in} is

$$H_{in} = C_w R_{in} \left(T_{in} - \frac{T_1 + T_2}{2} \right)$$

(18)

where T_{in} is the mean temperature of the mass flux into the control volume, $(T_1 + T_2)/2$ the mean temperature of the control volume under consideration, and C_w the specific heat of pure water at 0 °C.

The heat flux exported from the ice forming layer into the laminar layer is

$$H_{ifl-wf} = I_0 L_f + C_w R_{wf-ifl} \left(\frac{T_1 + T_2}{2} - T_0 \right)$$

(19)

The bulk heat flux between the laminar and airstream is

$$H_{wf-a} = -k_w \frac{T_1 - T_2}{y_2 - y_1} - C_w R_{wf-ifl} \left(\frac{T_1 + T_2}{2} - T_2 \right)$$

(20)

where k_w is the thermal conductivity of water ($0.58 \ W \cdot m^{-1} \cdot K^{-1}$).

The energy balance for the outer surface of the laminar film is

$$H_c + H_a = 0$$

(21)

where H_a is the heat flux on the outer surface of the laminar film, and H_c the conductive heat flux directed through the laminar layer into airstream from the ice forming layer, which is a component of H_{wf-a}.

The conductive heat flux is

$$H_c = k_w \frac{T_1 - T_2}{y_2 - y_1}$$

(22)

H_a can be expressed as follows (more details can be found in Ref. [23]):

$$H_a = -h_{cv}(T_2 - T_a) - \frac{\varepsilon h_{cv} L_v}{c_p P_a} \left(\frac{Pr}{Sc} \right)^{0.63} (e_3 - RHe_a) - \sigma a(T_2 - T_a) \\ - C_w R_{imp}(T_2 - T_{imp})$$

(23)

where h_{cv} and L_v (2.26×10^6 J $\cdot$ kg^{-1}) are convective heat transfer coefficient and the specific latent heat of vaporization respectively, e_3 and e_a represent the saturation vapor pressures at temperatures T_a and T_3 respectively; σ is the Stefan–Boltzman constant (5.67×10^{-8}), ε the ratio of the molecular weights of water and dry air (0.622), and r_a linearization constant for thermal radiation (8.1×10^7). RH is the relative humidity of the air, Sc is Schmidt number (0.7), and Pr is Prandtl number.

The definition of ice fraction can be written as

$$f = \frac{I_0}{I_0 + R_0} \qquad (24)$$

k is the function of k_w and k_i:

$$k = k_i f + k_w (1 - f) \qquad (25)$$

where k_i is the thermal conductivity of ice (2.25 W $\cdot$ m^{-1} $\cdot$ K^{-1}).

ICING REGIMES

Once the thermal profile of surficial layers is known, the growth regime can be determined. Conventionally, ice shapes are generally classified as glaze, mixed and rime accretions. As air temperature rises, the spongy icing model identifies icing conditions that include the rime accretion, the spongy without water film, the spongy with water film, and the glaze accretion. The outer surface temperature of the laminar layer, T_2 is determined by the heat balance equation for the outer surface of the laminar layer. If the calculated flux of entrapped liquid satisfies the condition, $R_0 = 0$, then the glaze icing regime will occur, i.e. both f and R_0 are equal to zero. The spongy with water film regime is predicted if the impinging mass flux of water is large enough to form both a water film of excess liquid and a spongy accretion, i.e. $R_{out} > 0$. As the temperature drops, the formation rate of pure ice becomes larger (Eq. (13)). If the impinging flux, R_{imp} is large enough to form a spongy accretion with no excess liquid ($R_{out} = 0$), and the formation rate of pure ice is less than the impinging flux (i.e. $I_0 < R_{imp}$ but $I_0 + R_0 > R_{imp}$), in this case the spongy without water film regime occurs. If T_2 is below 0 °C and the formation rate of pure ice is more than the flux of impinging droplets, no liquid will be entrapped and the ice growth rate can be determined by the flux of impinging droplets, and rime regime occurs.

FURTHER STUDY OF SPONGY ICING MODEL

In Messinger icing model, one assumption is that the temperature of interface between the liquid and the ice is 0 °C, and the interface is infinitely small. Messinger model can predict rime ice, mixed ice and glaze ice, when glaze ice occurs, the surface temperature equals 0 °C. But in the present model the spongy ice layer, the icing interface and the water film must be supercooled in order to drive the conductive heat away from the icing interface towards the airstream, which has been concluded in experiments.[13] When there is no water entrapped in the spongy ice layer, the ice fraction equals 1. In rime and glaze cases, the ice fraction equals 1. For the present model, the growth rate of the ice matrix will not be less than the rate of the supercooled liquid through the icing interface:

$$V_0 = \frac{I_0}{\rho_i} + \frac{R_0}{\rho_w} \geqslant V_c = a(\Delta T)^b = \frac{I_0 + R_0}{\rho_w} \qquad (26)$$

In both rime or glaze ice cases, $R_0 = 0$ and we can substitute Eq. (11) into Eq. (10):

$$T_0 - T_1 = \frac{q_1(y_1 - y_0)}{2k} = \frac{(I_0 L_f - C_w I_0 \Delta T)(y_1 - y_0)}{2k} \qquad (27)$$

Substituting Eqs. (15) and (16) into Eq. (27):

$$I_0 = \frac{2k\Delta T}{k_r(c + d/\Delta T)(L_f - C_w \Delta T)} \qquad (28)$$

Substituting Eq. (28) into Eq. (26), we have the following equation:

$$-C_w c\Delta T^b + (L_f c - C_w d)\Delta T^{b-1} + L_f d\Delta T^{b-2} - \frac{2k}{ak_r \rho_i} \leqslant 0 \qquad (29)$$

Then, we can solve the above equation. For example, when $t_a = -4.4$ °C, it can be solved as $\Delta T \leqslant 0.743$ °C, which means when $t_a = -4.4$ °C, glaze regime occurs, and the icing interface temperature will be hold in this range $0 > T_1 \geqslant -0.743$ °C.

RESULTS

In Fig. 4 and Fig. 5, the water collection efficiency β is plotted vs surface distance s in mm. The surface distance is measured from the nose (a reference point where $s = 0$ mm) corresponding to the location where y equals zero. For the clean airfoil, the nose is at the leading edge; for the iced airfoils, the nose is located between the ice horns. Note that negative surface distance corresponds to the upper surface of the airfoil.

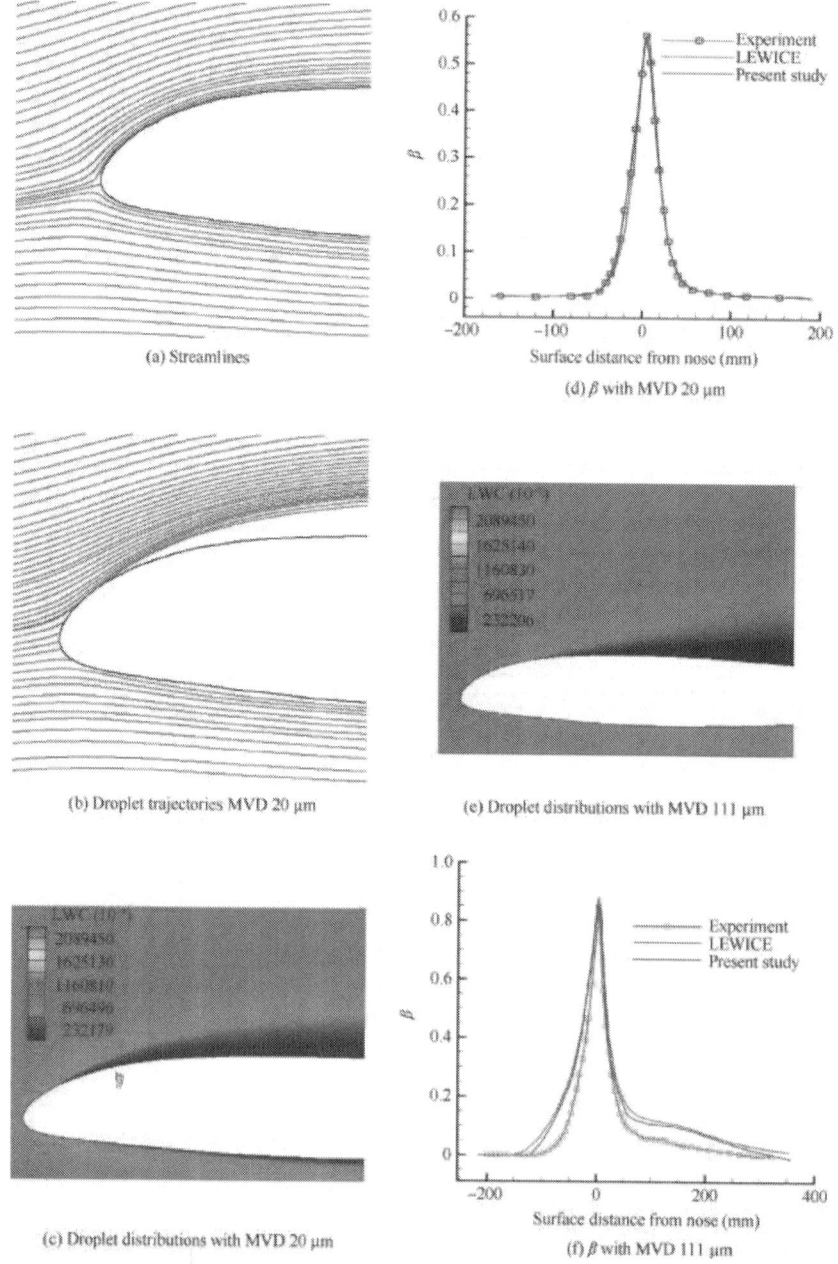

(a) Streamlines

(d) β with MVD 20 μm

(b) Droplet trajectories MVD 20 μm

(e) Droplet distributions with MVD 111 μm

(c) Droplet distributions with MVD 20 μm

(f) β with MVD 111 μm

Figure 4: Numerical and experimental results for the clean NACA23012 airfoil.

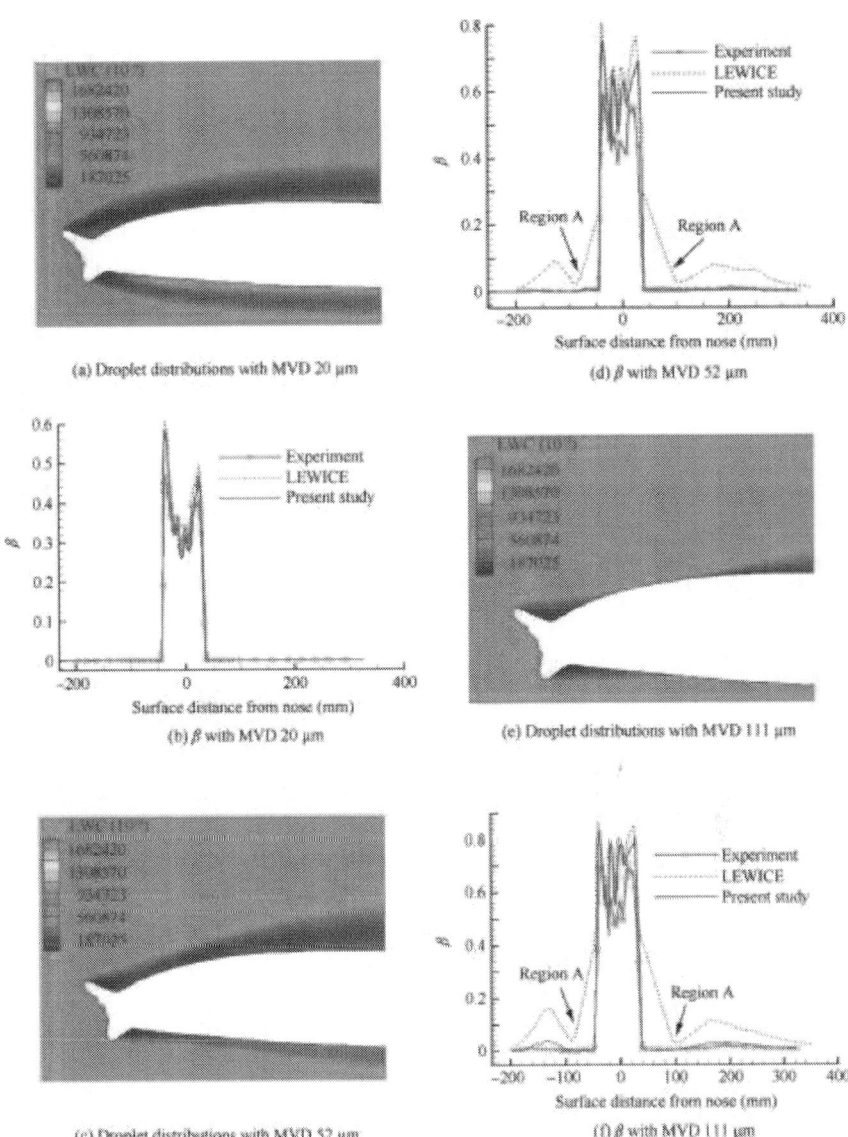

(a) Droplet distributions with MVD 20 μm

(b) β with MVD 20 μm

(c) Droplet distributions with MVD 52 μm

(d) β with MVD 52 μm

(e) Droplet distributions with MVD 111 μm

(f) β with MVD 111 μm

Figure 5: Numerical and experimental results for the iced NACA23012 airfoil.

Analysis of small and large droplets impingement data for all geometries tested are presented in this section. The analysis impingement data are obtained with present code and LEWICE. Fig. 4, Fig. 5 and Fig. 6 show the analysis data for the clean NACA23012 airfoil and the 22.5 min iced NACA23012 airfoil at 2.5° of AOA. Also shown are the experimentally measured values reported in Ref.[16].

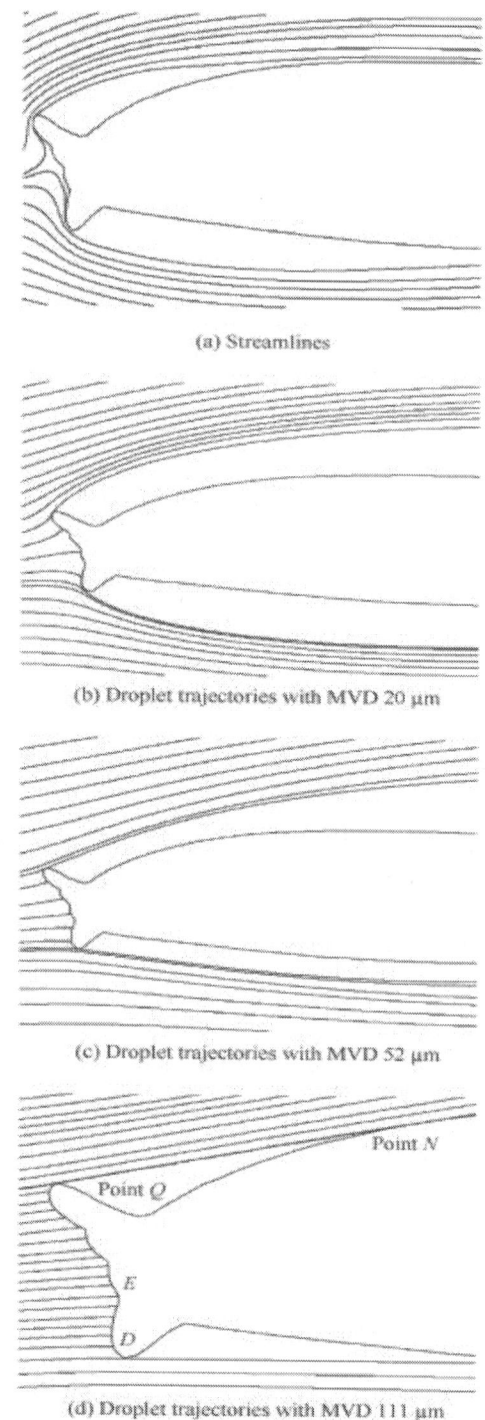

(a) Streamlines

(b) Droplet trajectories with MVD 20 μm

(c) Droplet trajectories with MVD 52 μm

Point N

Point Q

E

D

(d) Droplet trajectories with MVD 111 μm

Figure 6: Streamlines and droplet trajectories for iced NACA23012.

Impingement on Clean NACA23012 Airfoil

A Lagrangian particle tracking code usually computes droplets trajectories and uses them to calculate the water collection efficiency. However, for the Eulerian approach, the trajectories are not required for the computation of the collection efficiency. However, from the trajectories, one can tell the impingement limits easily. Fig. 4(a) and (b) demonstrate the streamlines and droplet trajectories for the clean NACA23012 airfoil with MVD = 20 μm. From the streamlines and the trajectories, it can be seen that droplets hit the leading edge due to the inertia, while the flow streamlines have the tendency to avoid the droplets hitting on the airfoil. For the case of MVD = 20 μm, the agreement between the experimental and CFD results is good. For the case of larger MVD = 111 μm, however, both the present code and the LEWICE code predict greater impingement limits and higher water collection efficiency compared to the experimental data. The possible reason is that the present code and the LEWCIE code do not account for the splashing effect, which is very common especially when the MVD is greater than 40 μm. It should be noted that the collection efficiency solved by the Eulerian method is lower than the collection efficiency solved by the Lagrangian method, especially around the nose of the airfoil ($s = 0$).

Impingement on NACA23012 with 22.5 Min Glaze Ice Shape

For the case of the 22.5 min ice shape and the large MVDs (52 and 111 μm), the LEWICE results in the downstream region of the horns (Region A, Figs. 5(d) and (f)) show a gradual decrease in β (collection efficiencies) compared to a sharp drop in the experimental and the present study results. The reason for this is due to the interpolation scheme used in the Lagrangian method. Generally, the interpolation scheme is fine; however, it has difficulties for geometries with multiple impingement regions which can occur on multi-element wings, highly cambered wings, and complex ice shapes. For example, one trajectory tangent with the aft impingement limit of a forward impingement region (Point Q, Fig. 6(d)) and another trajectory represents the forward limit of the aft impingement region (Point N, Fig. 6(d)); the collection efficiency between the two regions should be zero. However, for the Lagrangian method, it is not zero because of the interpolation scheme.

From Fig. 4 and Fig. 5, the results from the present study and the LEWICE code indicate higher local collection efficiency and greater impingement limits than the experimental results. There are three reasons for this. First, there is difference between the actual and the computed flow field, particularly in the region between the horns. Second, the droplet splashing is not simulated in the present code and the LEWICE code. Third, the errors are associated with the experimental investigation.

Comments on Droplet Trajectories

The droplet trajectories for an iced airfoil are depicted in Fig. 6 to illustrate the trends of impingement distribution in the present study. The presented droplet trajectories are for the 22.5 min ice shape with the 20, 52, and 111 μm MVD spray clouds. All trajectory computations are performed with the present code. The droplet trajectories shown in Fig. 6(b) for the 20 μm case demonstrate considerable deflection in the vicinity of the ice shape. As MVD becomes larger, the deflection of trajectories becomes progressively smaller (as shown in Fig. 6(c)); when MVD comes to 111 μm, the trajectories are practically straight (Fig. 6(d)). The reason for this is when MVD becomes larger, the inertia of the droplets becomes bigger too, so it is hard for streamlines to avoid the droplets hitting on the airfoil. Multiple local impingement peaks were observed between the ice horns. Fig. 6(d) can be used to explain how these peaks form. For example, the collection efficiency will be relatively high near Point D on the lower ice horn, for the surface is nearly normal to the incoming droplets; when the droplets hit the surface between Points D and E, the efficiency decreases due to the local slope of the surface. The reason of the peaks in the experimental results is the same as that of the peaks in the numerical results. However, the water re-impingement due to the droplet splashing may be the other reason.

Ice Accretion on NACA0012

The computation is performed at four different temperatures. Fig. 7, Fig. 8, Fig. 9 and Fig. 10 show the ice shapes after six minutes of simulation from the LEWICE and the present code, which are compared to the experimental data provided by shin and bond.[4] Fig. 7 shows the predicted ice shape at −19.4 °C, the ice fraction, and the mass flux of the control volume. There is no water film on the surface, and no droplet is entrapped. This is a typical rime ice. The predicted ice shape agrees well with the measured ice shape. Since it is rime and it is combined with the same Eulerian code, the spongy icing model and Messinger Model predict the same shape. Fig. 8 shows the other type of dry regime, called spongy without water film. The ice fraction holds in (0, 1), and there is no mass flux flow out of the control volume. For the temperature is relatively low, the Messinger model and spongy icing model do not have much difference, but the LEWICE predicts greater volume than the present code does. For the case of −10.0 °C (Fig. 9), this is a kind of wet ice accretion regime, named spongy with water film. The ice fraction holds in (0, 1) too, which means the liquid is entrapped in the spongy ice. The ice shape predicted by the present code with spongy icing model agrees well at suction side of the airfoil but slightly over-predicted at pressure side; while for the Messinger model, the ice shape fits not so well. There are some differences in predicted ice shapes calculated with spongy icing model and Messinger model, which is mainly due to the difference of predicted convective heat transfer coefficient. When it comes to −4.4 °C (Fig. 10), the predicted ice shape is a typical wet regime, glaze ice. The spongy icing model predicts a horn on the upper surface, which is the same as the LEWICE code does.

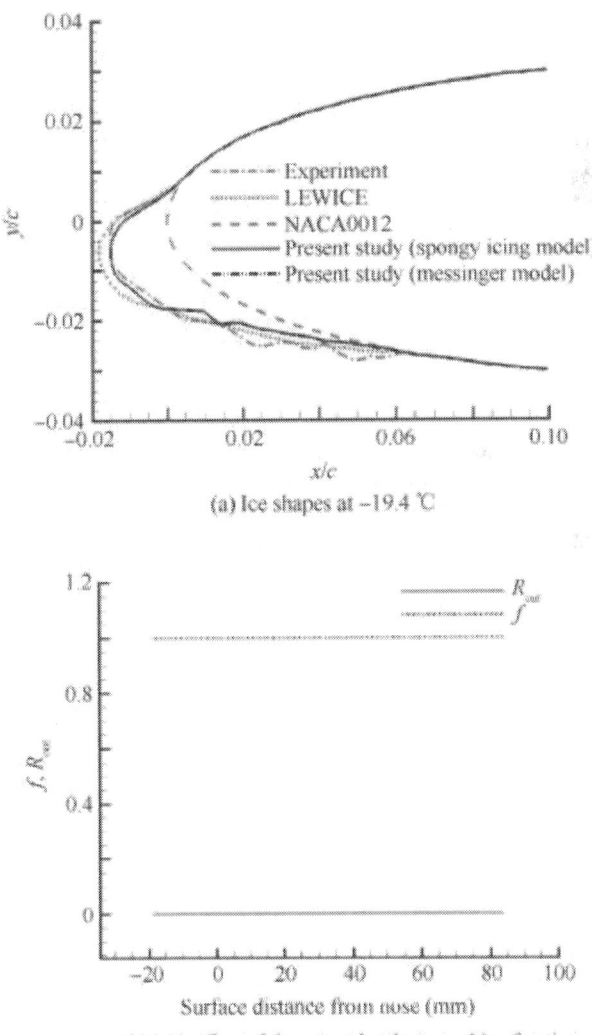

(a) Ice shapes at −19.4 °C

(b) Mass flux of the control volume and ice fraction

Figure 7: Comparison of numerical and experimental data at −19.4 °C.

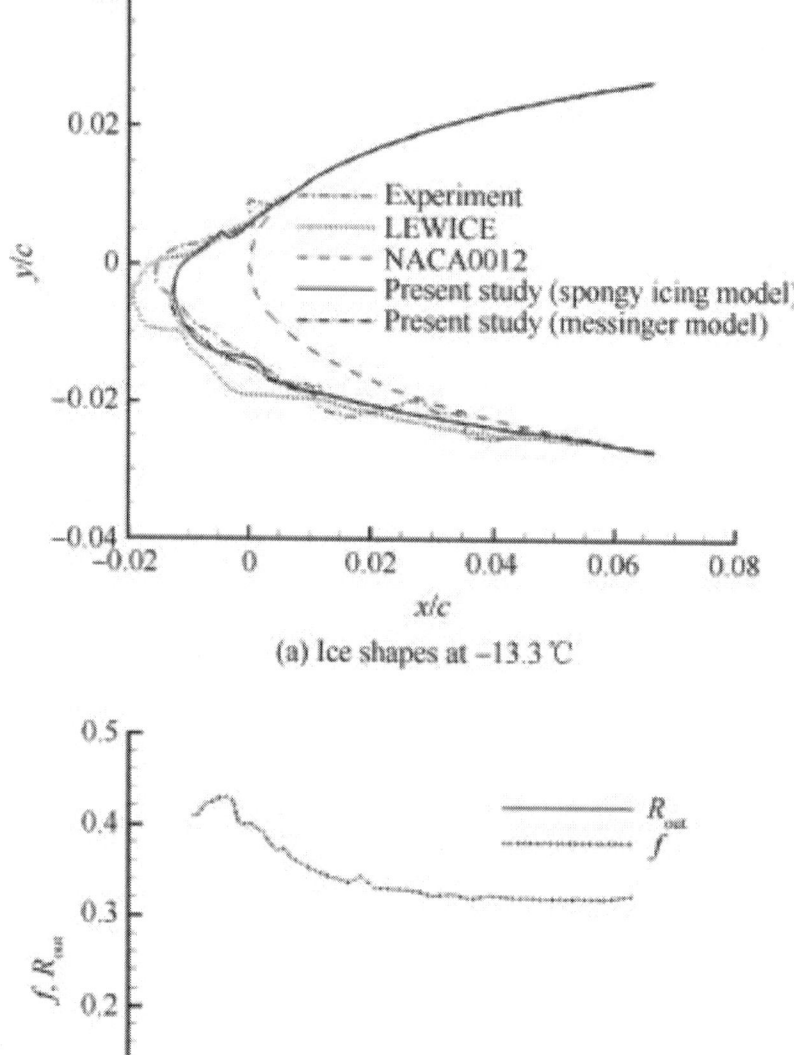

(a) Ice shapes at −13.3 ℃

(b) Mass flux of the control volume and ice fraction

Figure 8: Comparison of numerical and experimental data at −13.3 °C.

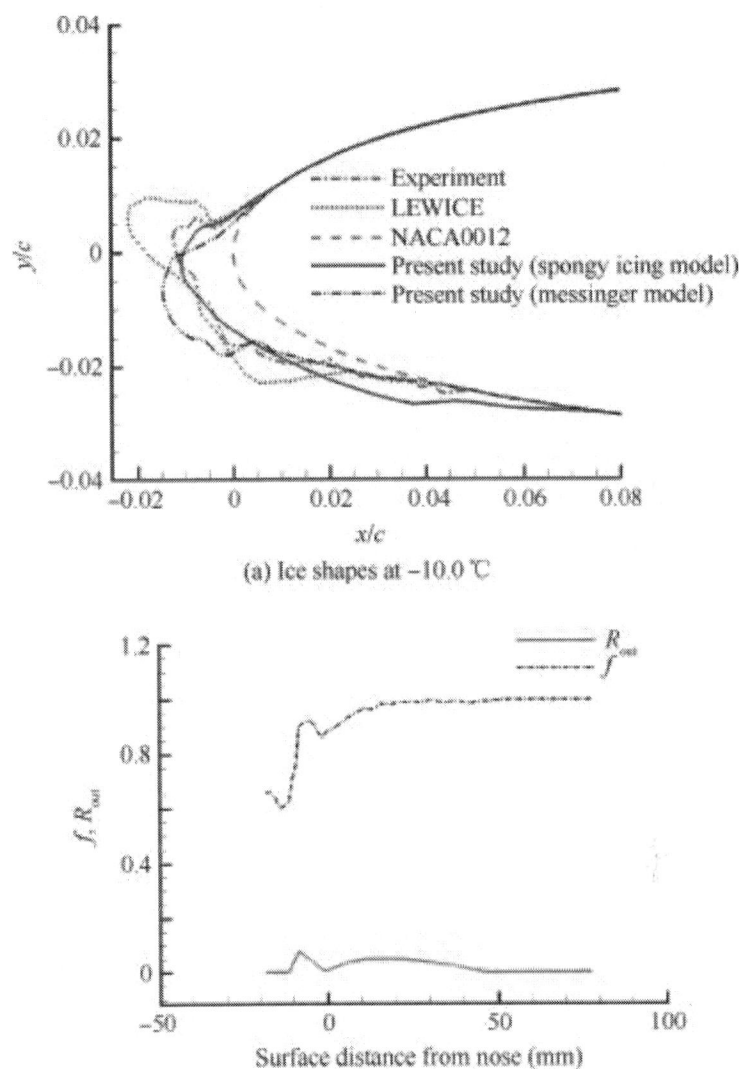

(a) Ice shapes at −10.0 ℃

(b) Mass flux of the control volume and ice fraction

Figure 9: Comparison of numerical and experimental data at −10.0 °C.

During the ice accretion process, it has been shown that the surface temperature is below the freezing point of water,[13] while it has been assumed that the temperature of the interface of ice and water film is at freezing point in Messinger model. Fig. 10(b) shows the surface temperature predicted at −4.4 °C by both models. The surface temperature predicted by Messinger model equals 0 °C while predicted by spongy icing model is below 0 °C; the temperature difference is up to 0.757 °C.

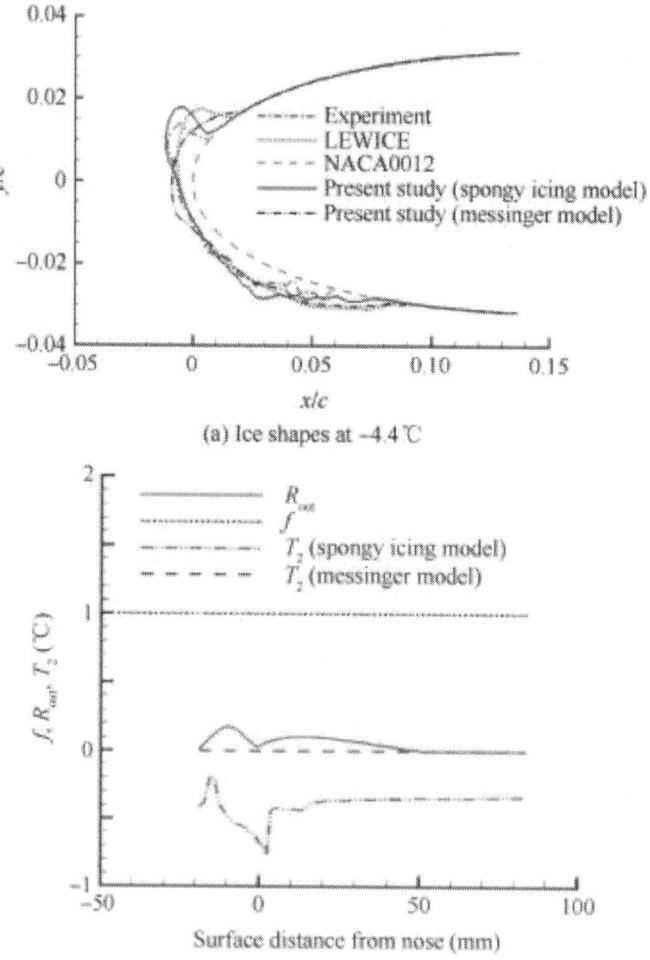

(a) Ice shapes at −4.4 ℃

(b) Mass flux of the control volume and ice fraction

Figure 10: Comparison of numerical and experimental data at −4.4 °C.

As noted above, for temperatures equal to −19.4 and −4.4 °C, corresponding to rime and glaze regime respectively. But for the other two cases (−13.3 and −10.0 °C), one is spongy without water film, another is spongy with water film.

Prediction of Ice Accretion and Surface Temperature on a Cylinder

The experimental results of Ref.[14] are chosen to validate the capability of the surface temperature prediction. During his experiment, in order to measure the surface temperature of a cylinder, a non-destructive remote sensing technique was employed. Test conditions are listed in Table 3. The ice shape predicted by the present code is similar to the measured shape (Fig. 11(a)). However, the thickness near the impingement limits is under-predicted and the ice thickness at stagnation point is over-predicted.

For the surface temperature shown in Fig. 11(b), both predicted and experimental results are below 0 °C. The predicted temperature is slightly higher than the experimental data, but the trend of the surface temperature is consistent.

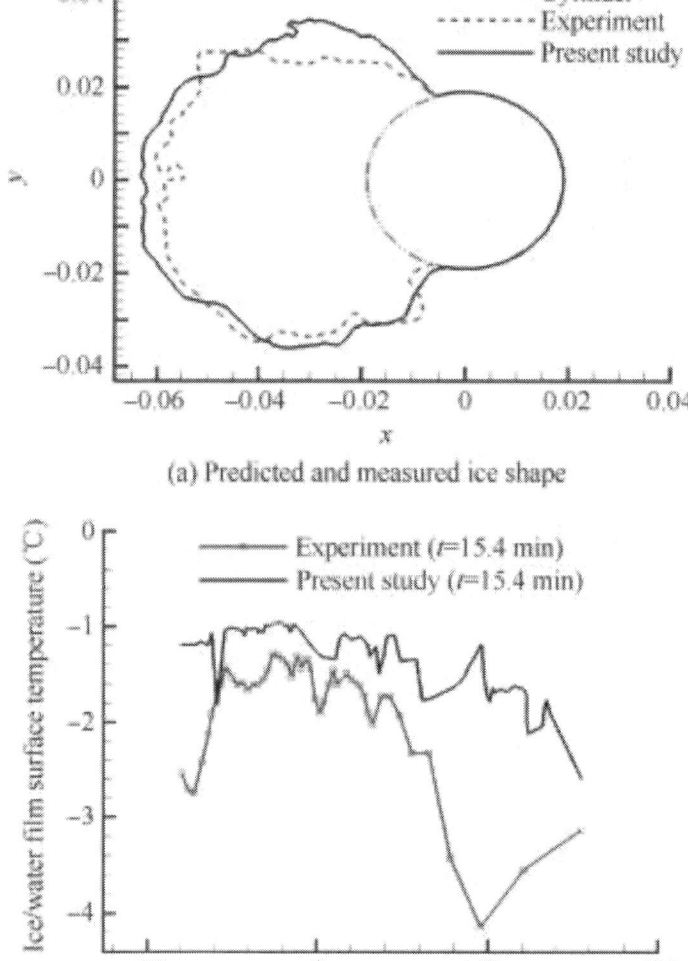

(a) Predicted and measured ice shape

(b) Predicted and measured temperature

Figure 11: Comparison of numerical and experimental data on cylinder.

CONCLUSION

A new spongy icing model based on Eulerian approach has been developed to describe glaze, spongy with water film, spongy without water film, and rime ice. A variety of experiments with different test conditions have been carried on in the NASA Lewis Icing Research Tunnel; the key parameters of the icing model are calibrated by these experimental data. The spongy icing model can be applied to aircraft icing condition, and it has a wide scope of application. The LWC range is 0–3.0 g/m^3, the MVD range is 8.0–156.0 μm, the velocity range is 0–200.0 m/s, the pressure range is 47.217–101.325 kPa and the temperature range is −30.0–0 °C.

To avoid numerical oscillations, a permeable wall condition is employed. Because of the numerical artifact caused by the interpolation scheme, the Eulerian approach is more feasible and accurate than the Lagrangian approach.

There are three differences between traditional models and the present icing model. The first one is the definition of the surface temperature. It has been assumed that the temperature of the interface between ice and water film is at the freezing point in traditional models, while in spongy icing model, the temperature of the interface is supercooled (e.g. when test temperature is −4.4 °C, the temperature difference is up to 0.757 °C, as shown in Fig. 10(b)). The second one is the capability of describing icing regimes. The present model can describe four different regimes while traditional models can only describe three at best. The last one is the capability of describing the sponginess of ice accretion which is observed by Fraser. The computational results are in good agreement with the experimental data. The results also point out the necessary of incorporating a droplet splash model into the present code.

ACKNOWLEDGMENTS

The authors thank professor Tom I-P Shih, head of school of Aeronautics and Astronautics, Purdue University, for giving so much help. Thanks are also due to Ph.D. Candidate Chien-Shing Lee and Research Scientist X. Chi from Purdue University for giving advices.

REFERENCES

1. Bourgault Y, Habashi WG, Dompierre J, Boutanios Z, Di Bartolomeo W. An Eulerian approach to supercooled droplets impingement calculations. AIAA-1997-0176; 1997.

2. Bragg MB. Rime ice accretion and its effects on airfoil performance [dissertation]. Columbus: The Ohio State University; 1981.

3. Shen XB, Lin GP, Bu XQ, Yu J, Hou PX. Three-dimensional analysis on ice shape of engine inlet lip. Acta Aeronaut et Asronaut Sin 2013;34(3):517–24 [Chinese].

4. Shin J, Bond T. Results of an icing test on a NACA0012 airfoil in the NASA Lewis Icing Research Tunnel. AIAA-1992-0647; 1992.

5. Bragg MB. Predicting rime ice accretion on airfoils. AIAA J 1985;23(3):381–7.

6. Potapczuk MG, Bidwell CS. Numerical simulation of ice growth on a MS-317 swept wing geometry. AIAA-1991-0263; 1991.

7. Fossati M, Khurram RA, Habashi WG. An ALE mesh movement scheme for long-term in-flight ice accretion. Int J Numer Methods Fluids 2012;68(8):958–76.

8. Messinger BL. Equilibrium temperature of an unheated icing surface as a function of air speed. J Aeronaut Sci 1953;20(1):29–42.

9. Myers TG. An extension to the messenger model for aircraft icing. AIAA J 2001;39(2):211–8.

10. Myers TG, Charpin JPF. A mathematical model for atmospheric ice accretion and water flow on a cold surface. Int J Heat Mass Transfer 2004;47(25):5483–500.

11. Bourgault Y, Beaugendre H, Habashi WG. Development of a shallow water icing model in FENSAP-ICE. J Aircr 2000;37(4):640–6.

12. Otta SP, Rothmayer AP. A simple boundary-layer water film model for aircraft icing. AIAA-2007-0902; 2007. 13. List R. Physics of supercooling of thin water skins covering gyrating hailstones. J Atmos 1990;47(15):1919–25.

13. Karev AR, Masoud F, Kollar LE. Measuring temperature of the ice surface during its formation by using infrared instrumentation. Int J Heat Mass Transfer 2007;50(3):566–79.

14. Blackmore RZ, Lozowski EP. A theoretical spongy spray icing model with surficial structure. Atmos Res 1998;49(4):267–88.

15. Papadakis M, Rachman A, Wong SC, Yeong HW, Hung KE, Bidwell CS. Water impingement experiments on a NACA23012 airfoil with simulated glaze ice shapes. AIAA-2004-0565; 2004.

16. Karadag A, Ganjoo DK, Shih TIP. A density-based method for computing incompressible multiphase flows Part 1: formulation and solution algorithm. AIAA-2000-0459; 2000.

17. Chi X, Zhu B, Shih TIP, Addy HE, Choo YK. CFD analysis of the aerodynamics of a business-jet airfoil with leading-edge ice accretion. AIAA-2004-0560; 2004.

18. Tirmizi SH, Gill WN. Effect of natural convection on growth velocity and morphology of dendritic ice crystals. J Cryst Growth 1987;85(3):488–502.

19. Lock GSH, Foster IB. Experiments on the growth of spongy ice near a stagnation point. J Glaciol 1990;36(123):143–50.

20. Dukler AE, Bergelin OP. Characteristics of flow in falling liquid films. Chem Eng Prog 1952;48(11):557–63.

21. William BW, Adam R. Validation results for LEWICE 2.0. NASA/CR-1999-208690; 1999.

22. Szilder K, Lozowski EP, Gates EM. Modeling ice accretion on non-rotating cylinders-the incorporation of time dependence and internal heat conduction. Cold Regions Sci Technol 1987;13(2):177–91.

CITATION

Xin Li, Junqiang Bai, Jun Hua, Kun Wang, Yang Zhang, A spongy icing model for aircraft icing, Chinese Journal of Aeronautics, Volume 27, Issue 1, February 2014, Pages 40-51, ISSN 1000-9361, http://dx.doi.org/10.1016/j.cja.2013.12.004.

Index